WORKSHEETS

FOR CLASSROOM OR LAB PRACTICE

CHRISTINE VERITY

INTERMEDIATE ALGEBRA

ELEVENTH EDITION

MARVIN BITTINGER

Indiana University Purdue University Indianapolis

Addison-Wesley
is an imprint of

Reproduced by Addison-Wesley from electronic files supplied by the author.

Publishing as Pearson Addison-Wesley, 75 Arlington Street, Boston, MA 02116.

ISBN-13: 978-0-321-61374-5
ISBN-10: 0-321-61374-0

1 2 3 4 5 6 BB 14 13 12 11

Addison-Wesley
is an imprint of

www.pearsonhighered.com

Table of Contents

Name: Date:
Instructor: Section:

Chapter R REVIEW OF BASIC ALGEBRA

R.1 The Set of Real Numbers

Learning Objectives

a Use roster notation and set-builder notation to name sets, and distinguish among various kinds of real numbers.

b Determine which of two real numbers is greater and indicate which, using < and >; given an inequality like $a < b$, write another inequality with the same meaning; and determine whether an inequality like $-2 \leq 3$or $4 > 5$ is true.

c Graph inequalities on the number line.

d Find the absolute value of a real number.

Key Terms

Use the vocabulary terms listed below to complete each statement in Exercises 1–4.

inequality **opposite** **set-builder notation** **roster method**

1. The ___________________ of 6 is –6.

2. The solution $\{x \mid x \geq -3\}$ is written using ___________________ .

3. The solution $\{-3, -2, -1\}$ is written using ___________________ .

4. The symbol $\leq$ is used to form a(n) ___________________ .

Objective a Use roster notation and set-builder notation to name sets, and distinguish among various kinds of real numbers.

Given the numbers $\frac{3}{4}, -7, \sqrt{14}, 0, 9, \sqrt{36}, \pi, -\frac{2}{3} \ldots$:

5. Name the whole numbers. **5.** _______________

6. Name the natural numbers. **6.** _______________

7. Name the integers. **7.** _______________

8. Name the rational numbers. 8 ________________

9. Name the irrational numbers. 9. ________________

10. Name the real numbers. 10. ________________

Use roster method to name each set.

11. The set of all letters in the word "number" 11. ________________

12. The set of all negative integers greater than –6 12. ________________

13. The set of all odd natural numbers less than 9 13. ________________

Use set-builder notation to name each set.

14. {–8,–7,–6,–5,–4} 14. ________________

15. The set of real numbers less than 5 15. ________________

16. The set of real numbers greater than or equal to –12 16. ________________

Objective b Determine which of two real numbers is greater and indicate which, using < and >; given an inequality like $a < b$, write another inequality with the same meaning; and determine whether an inequality like $-2 \leq 3$or $4 > 5$ is true.

Use either < or > for □ to write a true sentence.

17. 6 □ –7 17. ________________

18. –2 □ –5 18. ________________

Name: Date:
Instructor: Section:

19. $0 \square 8$ **19.** ____________

20. $-11.1 \square -5.2$ **20.** ____________

21. $-17\frac{3}{4} \square \frac{17}{100}$ **21.** ____________

22. $\frac{8}{27} \square \frac{5}{17}$ **22.** ____________

Write a different inequality with the same meaning.

23. $x \geq -2$ **23.** ____________

24. $14\frac{3}{5} < y$ **24.** ____________

25. $w > 4.7$ **25.** ____________

Write true or false.

26. $-3 \geq -3$ **26.** ____________

27. $1 \leq -1$ **27.** ____________

28. $-17 \geq -15\frac{3}{4}$ **28.** ____________

Objective c Graph inequalities on the number line.

Graph each inequality.

29. $x \geq 3$ **29.**

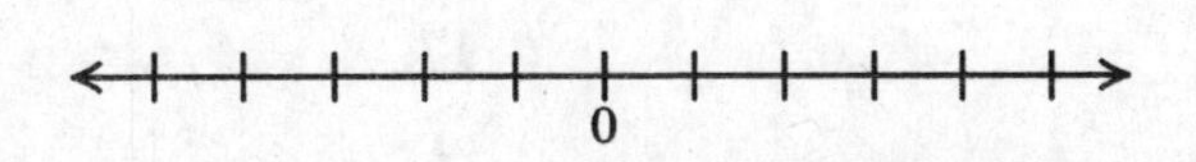

30. $x < -4$

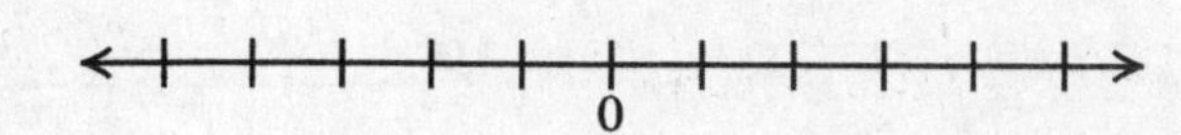

30. ______________

31. $x < 3$

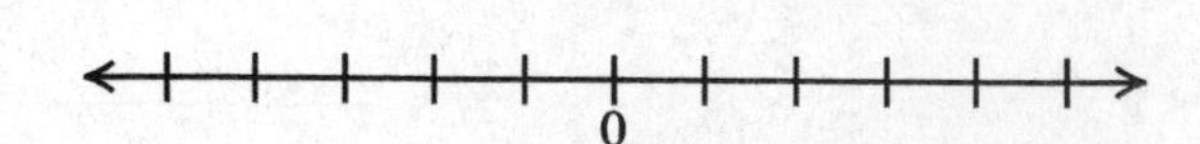

31. ______________

32. $x \geq 6$

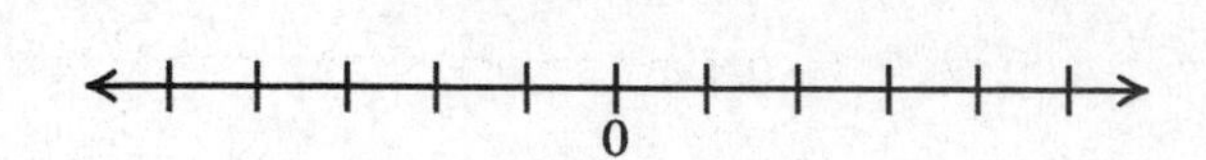

32. ______________

Objective d Find the absolute value of a real number.

Find the absolute value.

33. $|-8|$ 33. ______________

34. $|33|$ 34. ______________

35. $|-243|$ 35. ______________

36. $\left|-\frac{5}{8}\right|$ 36. ______________

37. $|18.9|$ 37. ______________

38. $|806|$ 38. ______________

39. $\left|\frac{0}{-6}\right|$ 39. ______________

Name: Date:
Instructor: Section:

Chapter R REVIEW OF BASIC ALGEBRA

R.2 Operations with Real Numbers

Learning Objective

a Add real numbers.
b Find the opposite, or additive inverse, of a number.
c Subtract real numbers.
d Multiply real numbers.
e Divide real numbers.

Key Terms

Use the vocabulary terms listed below to complete each statement in Exercises 1–4.

opposites **reciprocals** **additive** **multiplicative**

1. Two numbers whose sum is 0 are called ____________________, or ____________________ inverses, or each other.

2. Two numbers whose product is 1 are called ____________________, or ____________________ inverses, or each other.

Objective a Add real numbers.

Add.

3. $-12+(-9)$ **3.** ________________

4. $9+(-2)$ **4.** ________________

5. $-14+(-14)$ **5.** ________________

6. $6+(-15)$ **6.** ________________

7. $-33+11$ **7.** ________________

8. $-41+0$ **8.** ________________

9. $-5.4+6.9$ **9.** ________________

10. $-\frac{5}{8}+\frac{1}{8}$ **10.** ________________

11. $\frac{5}{6}+\left(-\frac{9}{10}\right)$ 11. ____________

12. $-\frac{2}{9}+\frac{1}{3}$ 12. ____________

Objective b Find the opposite, or additive inverse, of a number.

Evaluate –a for each of the following.

13. $a=0$ 13. ____________

14. $a=-11$ 14. ____________

15. $a=5.9$ 15. ____________

Find the opposite (additive inverse).

16. 14 16. ____________

17. 0 17. ____________

18. $-4y$ 18. ____________

Objective c Subtract real numbers.

Subtract.

19. $6-14$ 19. ____________

20. $17-17$ 20. ____________

21. $19-(-19)$ 21. ____________

22. $-24-24$ 22. ____________

23. $-5-(-5)$ 23. ____________

24. $-\frac{25}{4}-\left(-\frac{17}{4}\right)$ 24. ____________

25. $-\frac{3}{5}-\frac{7}{10}$ 25. ____________

Name: Date:
Instructor: Section:

26. $\frac{7}{8}-\frac{9}{16}$ 26. ______________

Objective d Multiply real numbers.

Multiply.

27. $-9(-4)$ 27. ______________

28. $-9\cdot 17$ 28. ______________

29. $(-1.6)(-8.5)$ 29. ______________

30. $-9(-13)(4)$ 30. ______________

31. $-\frac{135}{29}\cdot\left(\frac{-29}{135}\right)$ 31. ______________

32. $\left(-\frac{5}{8}\right)\left(-\frac{5}{8}\right)\left(-\frac{5}{8}\right)$ 32. ______________

33. $-6\left(-\frac{7}{6}\right)$ 33. ______________

34. $-\frac{15}{11}\cdot\left(-\frac{3}{2}\right)$ 34. ______________

Objective e Divide real numbers.

Divide, if possible.

35. $\frac{56}{-7}$ 35. ______________

36. $-42\div(-6)$ 36. ______________

37. $\frac{-90}{-1.5}$ 37. ______________

38. $\frac{-97}{0}$ 38. ______________

39. $\dfrac{0}{-13}$

39. ______________

40. $\dfrac{5}{7x-7x}$

40. ______________

41. $-\dfrac{15}{3}$

41. ______________

42. $\dfrac{-7.6}{-19}$

42. ______________

Find the reciprocal of each number.

43. $\dfrac{5}{4}$

43. ______________

44. 27

44. ______________

45. $-\dfrac{x}{y}$

45. ______________

46. $\dfrac{1}{7x}$

46. ______________

Divide.

47. $\dfrac{8}{5} \div \left(-\dfrac{7}{9}\right)$

47. ______________

48. $(-96.6) \div (-16.1)$

48. ______________

49. $720 \div (-0.12)$

49. ______________

50. $\dfrac{5}{6} \div \left(-\dfrac{7}{5}\right)$

50. ______________

51. $-5.5 \div 2.2$

51. ______________

52. $-51.1 \div (-7.3)$

52. ______________

53. $\dfrac{-13}{-9+9}$

53. ______________

Name: Date:
Instructor: Section:

Chapter R REVIEW OF BASIC ALGEBRA

R.3 Exponential Notation and Order of Operations

Learning Objectives

a Rewrite expressions with whole-number exponents, and evaluate exponential expressions.

b Rewrite expressions with or without negative integers as exponents.

c Simplify expressions using the rules for order of operations.

Key Terms

Use the vocabulary terms listed below to complete each statement in Exercises 1–2.

exponent **base**

1. In the expression 4^5, the 4 is the __________________.

2. In the expression 10^4, the 4 is the __________________.

Objective a Rewrite expressions with whole-number exponents, and evaluate exponential expressions.

Write exponential notation.

3. $7 \cdot 7 \cdot 7$ 3. ______________

4. $qqqq$ 4. ______________

5. $(4.2)(4.2)(4.2)(4.2)(4.2)$ 5. ______________

6. $\left(-\frac{3}{8}\right)\left(-\frac{3}{8}\right)\left(-\frac{3}{8}\right)$ 6. ______________

Evaluate.

7. $\left(\frac{1}{4}\right)^4$ — 7. ____________

8. $(-4.2)^2$ — 8. ____________

9. $(\sqrt{11})^1$ — 9. ____________

10. $(-96)^0$ — 10. ____________

11. $\left(\frac{3}{4}\right)^3$ — 11. ____________

12. $(0.2)^5$ — 12. ____________

Objective b Rewrite expressions with or without negative integers as exponents.

Rewrite using a positive exponent. Evaluate, if possible.

13. $\left(\frac{1}{3}\right)^{-4}$ — 13. ____________

14. $\left(\frac{5}{4}\right)^{-3}$ — 14. ____________

15. $\frac{1}{x^{-8}}$ — 15. ____________

16. $(-13)^{-1}$ — 16. ____________

Name: Date:
Instructor: Section:

Rewrite using a negative exponent.

17. $\frac{1}{5^4}$ **17.** ______________

18. $\frac{1}{a^{14}}$ **18.** ______________

19. $\frac{1}{(-15)^2}$ **19.** ______________

20. $\frac{1}{y^6}$ **20.** ______________

Objective c Simplify expressions using the rules for order of operations.

Simplify.

21. $(9+5)+6$ **21.** ______________

22. $80-(13+4)$ **22.** ______________

23. $120\div(20\div4)$ **23.** ______________

24. $(4+3)^3$ **24.** ______________

25. 6^2+9^2 **25.** ________________

26. $(36-24)^2-(16-8)^2$ **26.** ________________

27. $5\cdot 8-6$ **27.** ________________

28. $240\div 30+10$ **28.** ________________

29. $5(12-4)^2-3(1+6)^2$ **29.** ________________

Name: Date:
Instructor: Section:

Chapter R REVIEW OF BASIC ALGEBRA

R.4 Introduction to Algebraic Expressions

Learning Objectives
a Translate a phrase to an algebraic expression.
b Evaluate an algebraic expression by substitution.

Key Terms
Use the vocabulary terms listed below to complete each statement in Exercises 1–6.

algebraic expression	**constant**	**evaluating**
substituting	**value**	**variable**

1. A combination of letters, numbers, and operation signs, such as $16x-18y$ is called a(n) ____________________ .

2. A letter that can represent various numbers is a(n) ____________________ .

3. A letter that can stand for just one number is a(n) ____________________ .

4. When we replace a variable with a number, we are ____________________ for the variable.

5. When we replace all variables in an expression with numbers and carry out the operations, we are ____________________ the expression.

6. The results of evaluating an algebraic expression is called the ____________________ of the expression.

Objective a Translate a phrase to an algebraic expression.

Translate each phrase to an algebraic expression.

7. Eight more than some number **7.** ________________

8. Twenty less than some number

8. ________________

9. b divided by x

9. ________________

10. c subtracted from a

10. ________________

11. The product of two numbers

11. ________________

12. Six multiplied by some number

12. ________________

13. Ten more than seven times some number

13. ________________

14. Five less than the product of two numbers

14. ________________

15. Four times some number plus seven

15. ________________

16. The sum of twice a number plus five times another number

16. ________________

17. The price of a treadmill after a 25% reduction if the price before the reduction was p

17. ________________

18. Raena drove a speed of 55 mph for t hours. How far did Raena drive? (See Exercise 8.)

18. ________________

Name: Date:
Instructor: Section:

Objective b Evaluate an algebraic expression by substitution.

Evaluate.

19. $10z$, when $z = 3$

19. ________________

20. $\frac{x}{y}$, when $x = 42$ and $y = 6$

20. ________________

21. $\frac{5p}{q}$, when $p = 8$ and $q = 10$

21. ________________

22. $\frac{a-b}{3}$, when $a = 25$ and $b = 13$

22. ________________

Solve.

23. The area A of a triangle with base b and height h is given by $A = \frac{1}{2}bh$. Find the area when $b = 32$ cm (centimeters) and $h = 15$ cm.

23. 

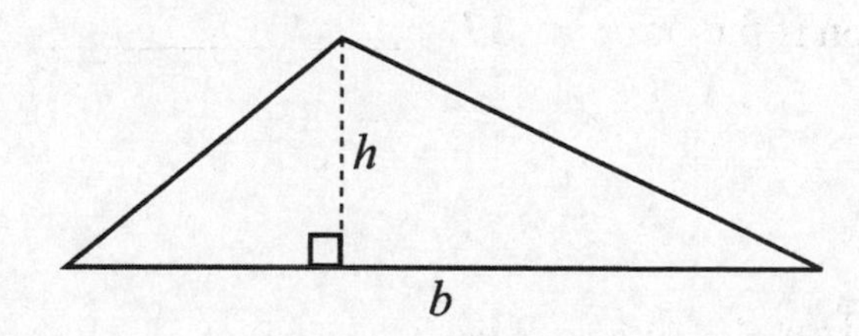

24. A driver who drives at a constant speed of r mph for t hr will travel a distance d mi given by $d = rt$ mi. How far will a driver travel at the speed of 70 mph for 3 hr?

24. ______________

25. The simple interest on a principal of P dollars at interest rate r for time t, in years, is given by $I = Prt$. Find the simple interest on a principal of \$1500 at 6% for 3 yr. (Hint: 6% = 0.06)

25. ______________

26. A rectangular piece of paper is 5 in. wide and 9 in. long. Find its area.

26. ______________

Name: Date:
Instructor: Section:

Chapter R REVIEW OF BASIC ALGEBRA

R.5 Equivalent Algebraic Expressions

Learning Objective

a Determine whether two expressions are equivalent by completing a table of values.
b Find equivalent fraction expressions by multiplying by 1, and simplify fraction expressions.
c Use the commutative laws and the associative laws to find equivalent expressions.
d Use the distributive laws to find equivalent expressions by multiplying and factoring.

Key Terms
In Exercises 1–5, match the term with the appropriate example in the column on the right.

1. _____ commutative law of addition

2. _____ associative law of multiplication

3. _____ identity property of 1

4. _____ commutative law of multiplication

5. _____ associative law of addition

a) $a \cdot 1 = 1 \cdot a$

b) $ab = ba$

c) $a + b = b + a$

d) $a + (b + c) = (a + b) + c$

e) $a \cdot (b \cdot c) = (a \cdot b) \cdot c$

Objective a Determine whether two expressions are equivalent by completing a table of values.

Complete the table by evaluating each expression for the given values. Then look for expressions that are equivalent.

6.

Value	$6(x-5)$	$6x-5$	$6x-30$
$x=-1$			
$x=5.2$			
$x=0$			

6. ________________

Objective b Find equivalent fraction expressions by multiplying by 1, and simplify fraction expressions.

Use multiplying by 1 to find an equivalent expression with the given denominator.

7. $\frac{9}{8}$; $8x$ **7.** ______________

8. $\frac{2}{3}$; $6x$ **8.** ______________

Simplify.

9. $\frac{75x}{40x}$ **9.** ______________

10. $\frac{-130b}{26b}$ **10.** ______________

Objective c Use the commutative laws and the associative laws to find equivalent expressions.

Use a commutative law to find an equivalent expression.

11. $x+4$ **11.** ______________

12. ay **12.** ______________

13. $ax+b$ **13.** ______________

Name: Date:
Instructor: Section:

Use an associative law to find an equivalent expression.

14. $x+(y-3)$ **14.** ______________

15. $(9\cdot a)\cdot b$ **15.** ______________

Use the commutative and associative laws to find three equivalent expressions.

16. $(x+y)+7$ **16.** ______________

17. $8\cdot(x\cdot y)$ **17.** ______________

Objective d Use the distributive laws to find equivalent expressions by multiplying and factoring.

Multiply.

18. $4(d+3)$ **18.** ______________

19. $5a(b-2c-4d)$ **19.** ______________

List the terms of each of the following.

20. $5a-7b+13$ **20.** ______________

Factor.

21. $9a+18b$ **21.** ______________

22. $20p+10q-15$ **22.** ______________

23. $\frac{3}{4}ab+\frac{1}{4}b^2$ **23.** ______________

24. $4x+12y-4$ **24.** ______________

25. $xz+z$ **25.** ______________

Name: Date:
Instructor: Section:

Chapter R REVIEW OF BASIC ALGEBRA

R.6 Simplifying Algebraic Expressions

Learning Objective
a Simplify an expression by collecting like terms.
b Simplify an expression by removing parentheses and collecting like terms.

Objective a Simplify an expression by collecting like terms.

Collect like terms.

1. $16w + 14w$ **1.** ______________

2. $5x^2 + x^2$ **2.** ______________

3. $12s - 19s + 18s$ **3.** ______________

4. $8u + 20 - 20u + 45$ **4.** ______________

5. $1.7x + 2.4y - 0.44x - 0.53y$ **5.** ______________

6. $\frac{3}{4}x + \frac{5}{6}y - 13 - \frac{2}{3}x - \frac{11}{6}y$ **6.** ______________

7. $-\frac{1}{3}a - \frac{1}{2}a + \frac{1}{3}b + \frac{1}{2}b - 29$ **7.** ______________

8. $a - 8a$ **8.** ______________

Objective b Simplify an expression by removing parentheses and collecting like terms.

Find an equivalent expression without parentheses.

9. $-(a + b + c)$ **9.** ______________

10. $-\left(-1.3x + 32.7y + 66z - \frac{1}{3}\right)$ **10.** ______________

Simplify by removing parentheses and collecting like terms.

11. $82t-(43t+90)$ **11.** ______

12. $-8(y+2)-7(y+9)$ **12.** ______

13. $7x-8-(10-4x)$ **13.** ______

14. $-6t+(5t-13)-3(2t+5)$ **14.** ______

Simplify.

15. $3\{[3(z-4)+17]-[2(5z-3)+3]\}$ **15.** ______

16. $4\{-3+5[6-3(5+7)]\}$ **16.** ______

17. $[5(x+7)-11]-[3(x-9)+12]$ **17.** ______

18. $5\{[6(x-2)+3^2]-4[3(x+7)-8^2]\}$ **18.** ______

19. $-\frac{1}{3}(12t-w)+\frac{1}{2}(-18t+6)+7$ **19.** ______

20. $8x-\{6[5(2x-9)-(7x-41)]+14\}$ **20.** ______

21. $8\{-8+9[6-4(3+5)]\}$ **21.** ______

Name: Date:
Instructor: Section:

Chapter R REVIEW OF BASIC ALGEBRA

R.7 Properties of Exponents and Scientific Notation

Learning Objectives

a Use exponential notation in multiplication and division.

b Use exponential notation in raising a power to a power, and in raising a product or a quotient to a power.

c Convert between decimal notation and scientific notation, and use scientific notation with multiplication and division.

Key Terms

Use the vocabulary terms listed below to complete each statement in Exercises 1–2.

exponential notation **scientific notation**

1. The expression 4^5 is written in ________________.

2. The expression 1.3×10^8 is written in ________________.

Objective a Use exponential notation in multiplication and division.

Multiply and simplify.

3. $u^4 \cdot u^9$ **3.**________________

4. $3^3 \cdot 3^0$ **4.**________________

5. $(5s)^9 \cdot (5s)^7$ **5.**________________

6. $\left(a^7b^8\right)\left(a^5b^6\right)$ **6.**________________

7. $x^4\left(xy^2\right)\left(xy\right)$ **7.**________________

Divide and simplify.

8. $\dfrac{u^8}{u^4}$

8.________________

9. $\dfrac{6^9 r^5}{6^5 r^3}$

9.________________

10. $\dfrac{3^5 s^5}{3^2 s^3}$

10.________________

11. $\dfrac{20m^6}{5m^3}$

11.________________

12. $\dfrac{x^8 y^7}{x^2 y^4}$

12.________________

Objective b Use exponential notation in raising a power to a power, and in raising a product or a quotient to a power.

Simplify.

13. $\left(3^2\right)^4$

13. ________________

14. $\left(a^5\right)^{-3}$

14. ________________

15. $\left(xy^2\right)^{-3}$

15. ________________

16. $\left(m^2 n^{-5}\right)^{-4}$

16. ________________

17. $\left(a^2 b^{-3} c^{-6}\right)^{-2}$

17. ________________

Name: Date:
Instructor: Section:

18. $\left(-5x^{4}y^{-3}\right)^{2}$ 18. ______________

19. $\left(\dfrac{x^{5}}{3}\right)^{2}$ 19. ______________

20. $\left(\dfrac{p^{2}q^{4}}{n}\right)^{5}$ 20. ______________

21. $\left(\dfrac{xy^{2}}{w^{3}z}\right)^{-3}$ 21. ______________

22. $\dfrac{11^{3a+2}}{11^{2a+2}}$ 22. ______________

23. $\dfrac{-15x^{a+1}}{5x^{3-a}}$ 23. ______________

24. $(7^{x})^{5y}$ 24. ______________

25. $(-10x^{-5}y^{6}z^{3})^{-4}$ 25. ______________

26. $\dfrac{30x^{a-b}y^{b-a}}{-6x^{a+b}y^{b+a}}$ 26. ______________

Objective c Convert between decimal notation and scientific notation, and use scientific notation with multiplication and division.

Convert each number to scientific notation.

27. 960,000,000,000 27. ______________

28. 0.00000000019 28. ______________

29. 770,000,000,000,000 29. ______________

30. 0.0000000067 30. ______________

Convert each number to decimal notation.

31. 4.21×10^{-7} **31.** ____________

32. 2.03×10^{11} **32.** ____________

33. 6.06×10^{-10} **33.** ____________

Multiply and write the answer in scientific notation.

34. $(1.5\times10^{3})(8.7\times10^{-5})$ **34.** ____________

35. $(9.4\times10^{-17})(7.4\times10^{20})$ **35.** ____________

36. $(9.8\times10^{-2})(8.37\times10^{-4})$ **36.** ____________

Divide and write the answer in scientific notation.

37. $\frac{9.2\times10^{-3}}{2.3\times10^{17}}$ **37.** ____________

38. $\frac{4.6\times10^{-16}}{7.2\times10^{16}}$ **38.** ____________

39. $\frac{3.9\times10^{14}}{8.4\times10^{-3}}$ **39.** ____________

Write the answer in scientific notation.

40. The distance light travels in 1 year (365 days) is approximately 5.87×10^{12} miles. How far does light travel in 75 weeks? **40.** ____________

41. A ream of a certain brand of paper weighs about 4.763 pounds. A ream contains 500 sheets of paper. How much does a sheet of paper weigh? **41.** ____________

Name: Date:
Instructor: Section:

Chapter 1 SOLVING LINEAR EQUATIONS AND INEQUALITIES

1.1 Solving Equations

Learning Objectives

a Determine whether a given number is a solution of a given equation.
b Solve equations using the addition principle.
c Solve equations using the multiplication principle.
d Solve equations using the addition principle and the multiplication principle together, removing parentheses where appropriate.

Key Terms

Use the vocabulary terms listed below to complete each statement in Exercises 1–8.

no solution	**identity**	**inverse**	**multiplication principle**
addition principle	**equation**	**solution**	**infinitely many solutions**

1. A(n) ____________________ is a number sentence that says that the expressions on either side of the equals sign represent the same number.

2. Any replacement for the variable that makes an equation true is called a(n) ____________________ of the equation.

3. The ____________________ states that for any real numbers a, b, and c, $a = b$ is equivalent to $a + c = b + c$.

4. The ____________________ states that for any real numbers a, b, and c, $c \neq 0, a = b$ is equivalent to $a \cdot c = b \cdot c$.

5. The multiplicative ____________________ of 3 is $\frac{1}{3}$.

6. The multiplicative ____________________ is 1 since $1 \cdot x = x$.

7. When solving an equation, if we end with a true equation, the equation has __________.

8. When solving an equation, if we end with a false equation, the equation has __________.

Objective a Determine whether a given number is a solution of a given equation.

Determine whether the given number is a solution of the given equation.

9. $-3;\ 4x-6=6$ **9.** ____________

10. $39;\ \frac{-x}{3}=-13$ **10.** ____________

11. $-6;\ 8-3x=26$ **11.** ____________

12. $14;\ 9+x=x+9$ **12.** ____________

13. $25;\ 6x+3=93$ **13.** ____________

Objective b Solve equations using the addition principle.

Solve using the addition principle. Don't forget to check.

14. $y+8=15$ **14.** ____________

15. $-24=x-19$ **15.** ____________

Name: Date:
Instructor: Section:

16. $-7 + z = 16$

16. ______________

17. $-49 + x = -71$

17. ______________

18. $r - 3.79 = 53.36$

18. ______________

Objective c Solve equations using the multiplication principle.

Solve using the multiplication principle. Don't forget to check.

19. $6x = 42$

19. ______________

20. $-9y = 126$

20. ______________

21. $-\frac{x}{7} = 88$

21. ______________

22. $-162 = -6z$

22. ______________

23. $5.7y = -45.6$

23. ________________

24. $-\frac{12}{5}t = \frac{3}{10}$

24. ________________

Objective d Solve equations using the addition principle and the multiplication principle together, removing parentheses where appropriate.

Solve using the principles together. Don't forget to check.

25. $8x + 9 = -79$

25. ________________

26. $-\frac{4}{9}x + \frac{5}{9} = -3$

26. ________________

27. $\frac{5}{6}x + \frac{8}{12}x = \frac{54}{12}$

27. ________________

28. $0.6t - 0.4t = 2.4$

28. ________________

29. $9x - 46 = 4x - 16$

29. ________________

Name: Date:
Instructor: Section:

30. $4x-15=33+4x$ **30.** ______________

31. $12t-16=1+7t$ **31.** ______________

32. $6-5a=a-2$ **32.** ______________

33. $5m-8=-8-6m-9m$ **33.** ______________

34. $3x+7=15-6x+x$ **34.** ______________

35. $-14+6x=6x-14$ **35.** ______________

36. $10y-4=9+10y$ **36.** ______________

37. $25=5(4t-3)$ **37.** ______________

38. $60(n-4)=180$

38. ______________

39. $9x-(4x-6)=46$

39. ______________

40. $5(2x-7)=6-(x-3)$

40. ______________

41. $-55x+42=2[9-4(8x-3)]$

41. ______________

42. $3[8-5(-4x+1)]=60x+9$

42. ______________

43. $\frac{1}{6}(12y+6)-17=-\frac{1}{3}(6y-12)$

43. ______________

44. $5[3(6-y)-4(8+3y)]-132=-6[5(6+3y)-3]$

44. ______________

Name: Date:
Instructor: Section:

Chapter 1 SOLVING LINEAR EQUATIONS AND INEQUALITIES

1.2 Formulas and Applications

Learning Objective
a Evaluate formulas and solve a formula for a specified letter.

Key Terms
Use the vocabulary terms listed below to complete each statement in Exercises 1–2.

evaluating **formula**

1. A(n) ____________________ is an equation relating two or more quantities.

2. When we replace the variables in an expression with numbers and calculate the result,

we are ____________________ the expression.

Objective a Evaluate formulas and solve a formula for a specified letter.

Solve for the given letter.

3. $A = gq$, for g **3.** ________________

4. $A = \dfrac{p+q+r}{3}$, for q **4.** ________________

5. $A = 4\pi r^2$, for r^2 **5.** ________________

6. $M = \dfrac{x-y}{2}$, for x **6.** ________________

7. $c = \dfrac{4k}{w}$, for w

7. ______________

8. $t = \dfrac{xy}{z}$ for z

8. ______________

9. $e = kry$, for y

9. ______________

10. $H = \dfrac{rs^2}{k}$, for r

10. ______________

11. $r = 3k + 3u$, for k

11. ______________

12. $2L + 2w + h = P$, for L

12. ______________

13. $K = B + Bvy$, for B

13. ______________

Name: Date:
Instructor: Section:

14. $Cx + Gy = J$, for x

14. ______________

15. $D = \frac{8}{3}(m - 52)$, for m

15. ______________

16. $K = \frac{7}{6}\pi p^8$, for p^8

16. ______________

17. $P = B + Bry$, for B

17. ______________

18. The cost for one month of Camden's cell phone, in dollars, is given by the formula $c = 45 + 0.1m$, where m is the number of text messages sent or received that month. How much was his cell phone bill for a month in which he sent or received 80 text messages?

18. ______________

19. Young's rule for determining the size of a child's medicine dosage is given by $c = \frac{ad}{a + 12}$, where a is the child's age and d is the usual adult dosage. (Warning: Do not apply this formula without consulting a physician.) The usual adult dosage of a medication is 190 mg. Find the dosage for a 4-yr old child. Round to the nearest mg.

19. ______________

20. The interval time, I, in minutes, between appointments is related to the total number of minutes T that a doctor spends with patients in a day, and the number of appointments N, by the formula: $I = 1.08\frac{T}{N}$. If a doctor wants an interval time of 19 minutes and wants to see 21 appointments per day, how many hours a day should the doctor be prepared to spend with patients? Round to the nearest hundredth of an hour.

20. ______________

21. The number of calories K needed each day by a moderately active woman who weighs w pounds and is h inches tall and is a years old can be estimated by the formula $K = 917 + 6(w + h - a)$. If Marcia is moderately active, weighs 127 pounds, is 65 inches tall, and is 50 years old, how many calories does she need per day?

21. ______________

22. One formula for projecting human birth weight is given by $P = 9.337da - 299$, where
P = the projected birth weight in grams
d = the diameter of the head in centimeters, and
a = the circumference of the abdomen in centimeters at 29 weeks, using ultrasound.
Use the formula to estimate the birth weight of a baby which has a head diameter of 7.5 cm and an abdominal circumference of 38.2 cm. Round to the nearest gram.

22. ______________

Name: Date:
Instructor: Section:

Chapter 1 SOLVING LINEAR EQUATIONS AND INEQUALITIES

1.3 Applications and Problem Solving

Learning Objectives	
a	Solve applied problems by translating to equations.
b	Solve basic motion problems.

Key Terms

Use the vocabulary terms listed below to complete each statement in Exercises 1–5.

check **familiarize** **solve** **state** **translate**

1. To ___________________ yourself with a problem, read it carefully, choose a variable to represent the unknown, and make a drawing.

2. To ___________________ a problem into mathematical language, write an equation.

3. To ___________________ an equation, find all replacements that make the equation true.

4. Always ___________________ the answer in the original problem.

5. As a final problem-solving step, ___________________ the answer to the problem clearly.

Objective a Solve applied problems by translating to equations.

Solve.

6. A 60-in. board is cut into two pieces. One piece is four times the length of the other. Find the lengths of the pieces.

6. ________________

7. In a recent year, New Hampshire had 117 women holding legislative office. This was 53 more than the number of women holding office in Maryland. How many women held legislative office in Maryland?
Source: U.S. Census Bureau

7. ______________

8. A total of 2 in. of precipitation was recorded in East Lake City on May 11 and 12. The amount recorded on May 11 was three times the amount recorded on May 12. How much was recorded on May 12?

8. ______________

9. The sum of three consecutive integers is 69. What are the numbers?

9. ______________

10. The sum of three consecutive even integers is 198. What are the integers?

10. ______________

11. A rectangle has a perimeter of 88 ft. The length is 2 ft more than twice the width. Find the dimensions of the rectangle.

11. ______________

Name: Date:
Instructor: Section:

12. Caitlyn paid \$54.40 for a sweater during a 15%-off sale. What was the regular price?

12. ______________

13. Carlee paid \$27.03, including 6% tax, for decorations for a party. What was the cost of the decorations before tax?

13. ______________

14. The second angle of a triangle is twice as large as the first angle. The third angle is 12° more than four times the first angle. How large are the angles?

14. ______________

15. The balance in Clayton's charge card account grew 3%, to \$669.50, in one month. What was his balance at the beginning of the month?

15. ______________

16. Craig left an 18% tip for a meal. The total cost of the meal, including the tip, was \$51.92. What was the cost of the meal before the tip was added?

16. ______________

Objective b Solve basic motion problems.

Solve.

17. A canoe moves at a rate of 7 km/h in still water. How long will it take the canoe to travel 16 km downriver (with the current) if the current moves at a rate of 1 km/h?

17. ________________

18. Chris swims at a speed of 2 mph in still water. The current in a river is moving at 0.5 mph. How long will it take Chris to swim 0.3 mile upriver against the current?

18. ________________

19. A commercial jet has been instructed to climb from its present altitude of 9000 ft to a cruising altitude of 27,000 ft. If the plane ascends at a rate of 1500 ft/min, how long will it take to reach its cruising altitude?

19. ________________

Name: Date:
Instructor: Section:

Chapter 1 SOLVING LINEAR EQUATIONS AND INEQUALITIES

1.4 Sets, Inequalities, and Interval Notation

Learning Objectives

a Determine whether a given number is a solution of an inequality.
b Write interval notation for the solution set or the graph of an inequality.
c Solve an inequality using the addition principle and the multiplication principle and then graph the inequality.
d Solve applied problems by translating to inequalities.

Key Terms

Use the vow3cabulary terms listed below to complete each statement in Exercises 1–4.

equivalent **graph** **inequality** **set-builder notation**

1. A(n) ____________ is a number sentence with <, >, ≤, or ≥ as its verb.

2. A(n) ____________ of an inequality is a drawing that represents its solutions.

3. The sentences $x+4<10$ and $x<6$ are ____________ since they have the same solution set.

4. The solution set $\{x \mid x>2\}$ is written using ____________ .

Objective a Determine whether a given number is a solution of an inequality.

Determine whether each number is a solution of the given inequality.

5. $x \le -6$ **5.**

a) 0 **a)** ____________

b) –3 **b)** ____________

c) –6 **c)** ____________

d) –9 **d)** ____________

e) –5.4 **e)** ____________

6. $x > 10$

a) 0

b) -12

c) 18

d) 12.7

e) 10

6.

a) ____________

b) ____________

c) ____________

d) ____________

e) ____________

Objective b Write interval notation for the solution set or the graph of an inequality.

Write interval notation for the given set.

7. $\{x \mid -3 < x < 6\}$

8. $x \geq 3$

9. $x < -4$

7. ____________

8. ____________

9. ____________

Objective c Solve an inequality using the addition principle and the multiplication principle and then graph the inequality.

Solve and graph.

10. $x + 6 > 3$

10. ____________

____

0

Name: Date:
Instructor: Section:

11. $2x+4 \le x+7$

11.________________

____

0

12. $-12x > -36$

12.________________

____

0

13. $3x \ge 18$

13.________________

____

0

Solve.

14. $y+\frac{2}{7} \le \frac{7}{14}$

14.________________

____

15. $-10z+5 > 5-11z$

15.________________

____

16. $9x \ge -8$

16.________________

____

17. $\frac{-4}{7} > -6x$

17.________________

____

18. $6 + 9n < -21$

18.________________

____

19. $8x - 8 \leq 32$

19.________________

____

20. $9x - 9 < -54$

20.________________

____

21. $5x + 2 - 4x \leq 13$

21.________________

____

22. $6 - 20x \leq 5 - 11x - 8x$

22.________________

____

23. $\frac{x}{4} - 1 \leq \frac{1}{2}$

23.________________

____

Name: Date:
Instructor: Section:

24. $5(2x-3)<45$

24.______________

____

25. $5(2y-9)\geq 6(3y+4)$

25.______________

____

Objective d Solve applied problems by translating to inequalities.

Solve.

26. You are taking a math course in which there will be four tests, each worth 100 points. You have scores of 97, 94, and 97 on the first three tests. You must earn a total of 360 points in order to get an A. What scores on the last test will give you an A?

26.______________

27. In planning for a banquet, you find that one speaker charges $225 plus 50% of the total ticket sales. Another speaker charges a flat fee of $525. In order for the first speaker to produce more profit for your organization than the other speaker, what is the highest price you can charge per ticket, assuming that 200 people will attend?

27.______________

28. Most insurance companies will replace a vehicle any time an estimated repair exceeds 80% of its "blue book" value. Melissa's car had $6500 in repairs after an accident. What can be concluded about its "blue book" value?

28.________________

29. Carlos can be paid in one of two ways. Plan A is a salary of $500 per month, plus a commission of 8% of sales. Plan B is a salary of $653 per month, plus a commission of 5% of sales. For what amount of sales is Carlos better off selecting Plan A?

29.________________

Name: Date:
Instructor: Section:

Chapter 1 SOLVING LINEAR EQUATIONS AND INEQUALITIES

1.5 Intersections, Unions, and Compound Inequalities

Learning Objective

a Find the intersection of two sets. Solve and graph conjunctions of inequalities.
b Find the union of two sets. Solve and graph disjunctions of inequalities.
c Solve applied problems involving conjunctions and disjunctions of inequalities.

Key Terms

Use the vocabulary terms listed below to complete each statement in Exercises 1–5.

conjunction	**compound inequalities**	**disjunction**
union	**intersection**	

1. Two inequalities joined by the word "and" or the word "or" are called ____________.

2. A(n) ____________ is a compound sentence formed using the word "and."

3. The solution of a conjunction is the ____________ of the solutions sets of the individual sentences.

4. A(n) ____________ is a compound sentence formed using the word "or."

5. The solution of a disjunction is the ____________ of the solutions sets of the individual sentences.

Find the intersection or union.

6. $\{5,6,8,9\}\cap\{4,6,9,11\}$ 6.____________

7. $\{7,8,9,10,11\}\cup\{5,7,9,13\}$ 7.____________

8. $\{3,8,20,24\}\cup\{2,3,8,15\}$ 8.____________

9. $\{c,d,e,f,g\}\cap\{f,g,h,i,j\}$ 9.____________

10. $\{m, a, t, h\} \cap \{m, a, t\}$

10.________________

11. $\{2, 3, 5, 10\} \cup \varnothing$

11.________________

Objective a Find the intersection of two sets. Solve and graph conjunctions of inequalities.

Graph and write interval notation.

12. $-2 \leq x \leq 5$

12.________________

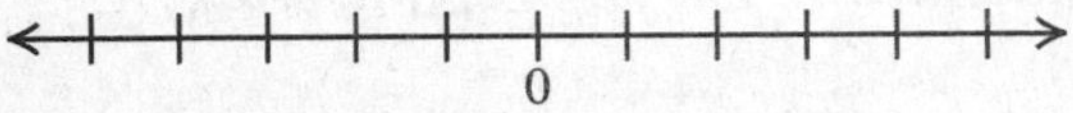

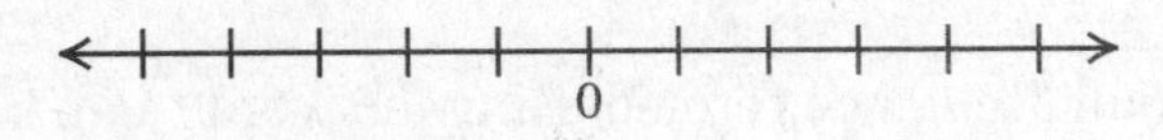

13. $6 \geq m$ *and* $m > -2$

13.________________

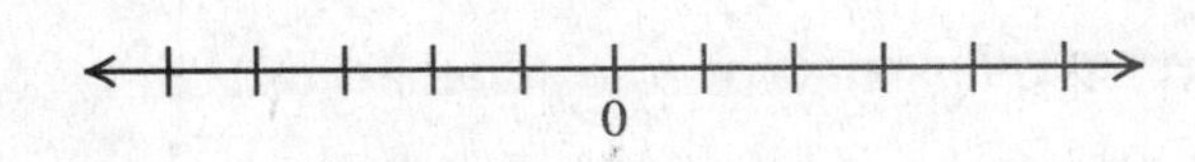

14. $x < 5$ *and* $x \geq 2$

14.________________

Solve and graph.

15. $-4 \leq x - 5 < 3$

15.________________

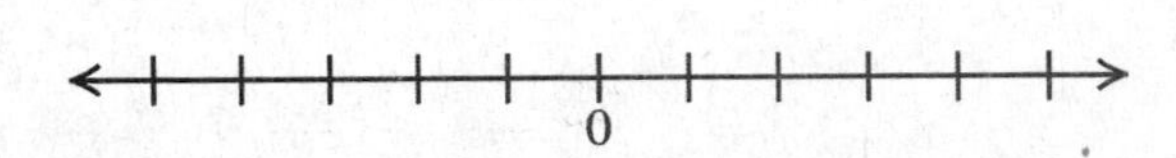

16. $-2 < 3t + 1$ *and* $2t - 5 \leq 7$

16.________________

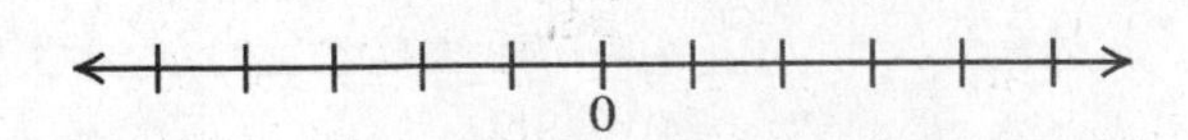

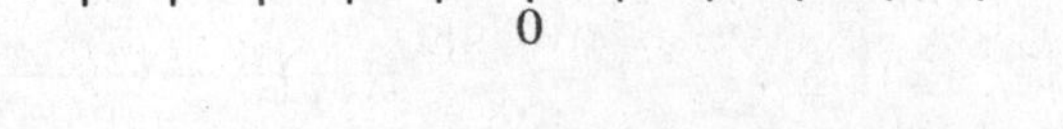

Name: Date:
Instructor: Section:

Solve.

17. $5 \geq \frac{x+1}{4} \geq -3$ **17.**______________

18. $-4 < 3x + 2 \leq 11$ **18.**______________

Objective b Find the union of two sets. Solve and graph disjunctions of inequalities.

Graph and write interval notation.

19. $t < 0$ *or* $t > 5$ **19.**______________

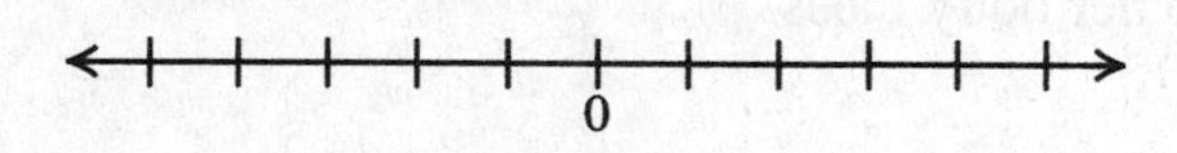

20. $y \geq -2$ *or* $-y \geq 3$ **20.**______________

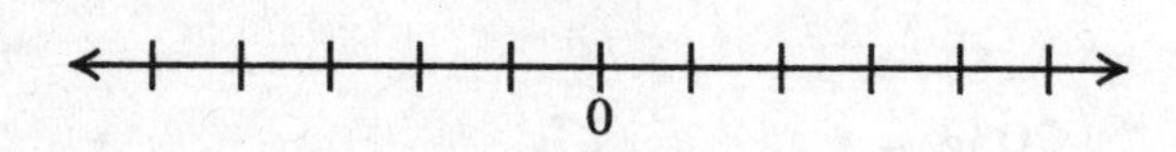

21. $x \geq -3$ *or* $x \geq -1$ **21.**______________

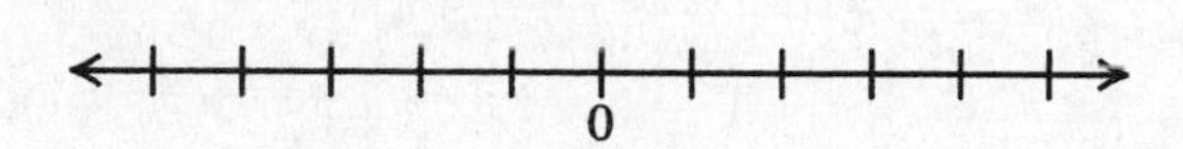

Solve and graph.

22. $2z + 3 \geq -5$ *or* $6z - 7 < 11$ **22.**______________

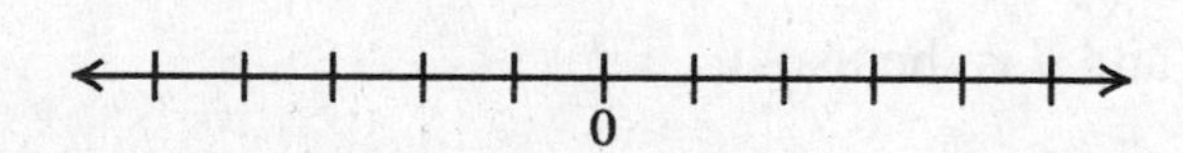

23. $3a - 5 \geq 8$ *or* $7 - 2a < 1$ **23.**______________

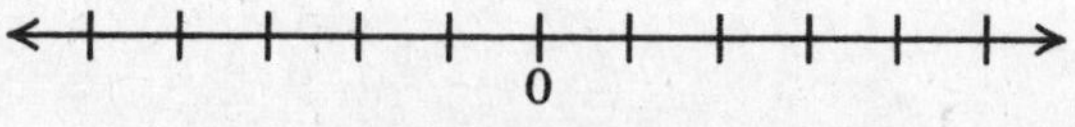

Solve.

24. $5x+2<-3$ or $5x-2>3$ **24.**______________

25. $-3m+4<10$ or $-3m-4\geq 5$ **25.**______________

Objective c Solve applied problems involving conjunctions and disjunctions of inequalities.

Solve.

26. Jen's height is 65 in. Use the formula $I=\dfrac{703w}{H^2}$ to find the weight w that will allow Jen to keep her body mass index I within the 18.5-24.9 range. **26.**______________

27. Jack's height is 74 in. Use the formula $I=\dfrac{703w}{H^2}$ to find the weight w that will allow Jack to keep his body mass index I within the 18.5-24.9 range. **27.**______________

28. The dosage of a medication for a 10-year old child must stay between 150 mg and 250 mg. Use the formula $c=\dfrac{ac}{a+12}$, where a is the child's age, and d is the usual adult dosage, to find the equivalent adult dosage. **28.**______________

Name: Date:
Instructor: Section:

Chapter 1 SOLVING LINEAR EQUATIONS AND INEQUALITIES

1.6 Absolute-Value Equations and Inequalities

Learning Objective

a Simplify expressions containing absolute-value symbols.
b Find the distance between two points on the number line.
c Solve equations with absolute-value expressions.
d Solve equations with two absolute-value expressions.
e Solve inequalities with absolute-value expressions.

Key Terms

Finish each statement in Exercises 1–5 with ***p***, ***–p***, or **0**, where *p* is a positive number.

1. The solutions of $|X| =$ ________ are those numbers that satisfy $X = p$ or $X = -p$.

2. The equation $|X| =$ ________ is equivalent to the equation $X = 0$.

3. The equation $|X| =$ ________ has no solution.

4. The solutions of $|X| < p$ are those numbers that satisfy ________ $< X <$ ________.

5. The solutions of $|X| > p$ are those numbers that satisfy $X <$ ________ *or* $X >$ ________.

Objective a Simplify expressions containing absolute-value symbols.

Simplify, leaving as little as possible inside absolute-value signs.

6. $|7x|$ **6.**________________

7. $|5x^2|$ **7.**________________

8. $|-6x^2|$ **8.**________________

9. $|-19z|$ 9. ____________

10. $\left|\frac{y^2}{-z}\right|$ 10. ____________

11. $\left|\frac{-8a^2}{4a}\right|$ 11. ____________

12. $\left|\frac{5x^2}{-15x}\right|$ 12. ____________

13. $\left|\frac{72x^5}{-8x}\right|$ 13. ____________

Objective b Find the distance between two points on the number line.

Find the distance between the points on the number line.

14. –11, –96 14. ____________

15. –56, 27 15. ____________

16. –2.6, –4.8 16. ____________

17. $\frac{3}{4}, -\frac{7}{12}$ 17. ____________

Name: Date:
Instructor: Section:

Objective c Solve equations with absolute-value expressions.

Solve.

18. $|x| = 4.3$ **18.**________________

19. $|t| = 0$ **19.**________________

20. $|x| = -18$ **20.**________________

21. $|4x - 5| = 8$ **21.**________________

22. $|x - 7| = 10$ **22.**________________

23. $|y - 5| = -15$ **23.**________________

24. $|3x - 7| = 4$ **24.**________________

25. $|5x| - 7 = 9$

25.________________

26. $\left|\frac{3-5t}{2}\right| = 11$

26.________________

Objective d Solve equations with two absolute-value expressions.

Solve.

27. $|2x+1| = |x-6|$

27.________________

28. $|x+6| = |x-7|$

28.________________

29. $|5t+4| = |2t-7|$

29.________________

Name: Date:
Instructor: Section:

30. $|3x-4|=|4-3x|$ **30.**________________

Objective e Solve inequalities with absolute-value expressions.

Solve.

31. $|x|\geq 3$ **31.**________________

32. $|x|<15$ **32.**________________

33. $|x+3|<7$ **33.**________________

34. $|x-3|+8\geq 14$ **34.**________________

35. $13-|2x+1|\leq 5$ **35.**________________

36. $|6-3a| \le 4$

37. $\left|\dfrac{7-2x}{3}\right| \ge \dfrac{5}{2}$

36.________________

37.________________

Name: Date:
Instructor: Section:

Chapter 2 GRAPHS, FUNCTIONS AND APPLICATIONS

2.1 Graphs of Equations

Learning Objectives

a Plot points associated with ordered pairs of numbers.
b Determine whether an ordered pair of numbers is a solution of an equation.
c Graph linear equations using tables.
d Graph nonlinear equations using tables.

Key Terms

Use the vocabulary terms listed below to complete each statement in Exercises 1–6.

axes **coordinates** **graph** **ordered pair** **origin** ***y*-intercept**

1. We graph number pairs on a plane using two perpendicular number lines called

__________________.

2. On the plane, the perpendicular number lines cross at a point called the

__________________.

3. The numbers in an ordered pair are called __________________.

4. The notation $(3,-2)$ is an example of a(n) __________________.

5. The __________________ of an equation is a drawing that represents all its solutions.

6. A graph crosses the y-axis at the __________________.

Objective a Plot points associated with ordered pairs of numbers.

Plot the following points.

7. $(2, 4)$ $(-3, 1)$ $(2,-5)$ $(-4,-1)$ $(0,-2)$ $(2, 0)$ $(0, 3)$ $(-4, 0)$

7. ______________

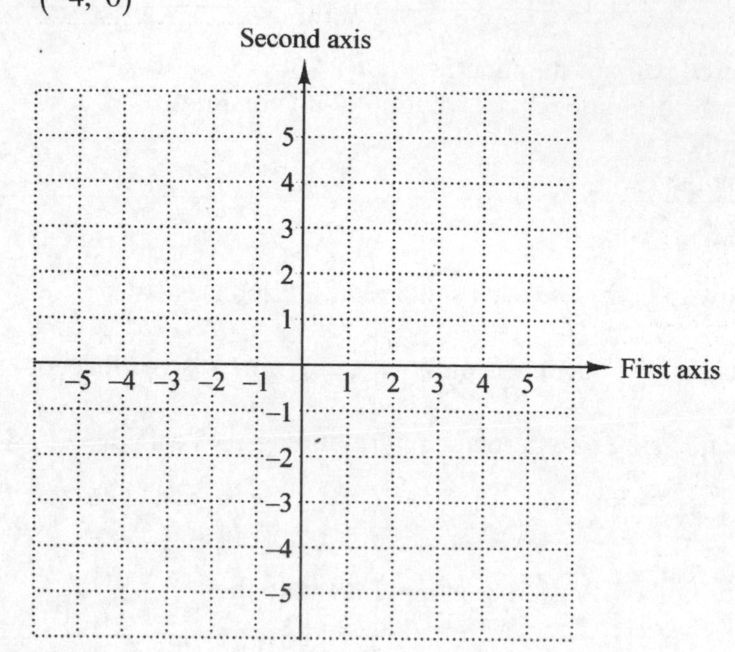

Objective b Determine whether an ordered pair of numbers is a solution of an equation.

Determine whether the given point is a solution of the equation.

8. $(1, 5)$; $2x - y = 3$

8. ______________

9. $(-2, 4)$; $3a + 2b = 2$

9. ______________

10. $(0,-3)$; $5x - y = 3$

10. ______________

Name: Date:
Instructor: Section:

In Exercises 11 and 12, an equation and two ordered pairs are given. Show that each pair is a solution of the equation. Then use the graph of the two points to determine another solution. Answers may vary.

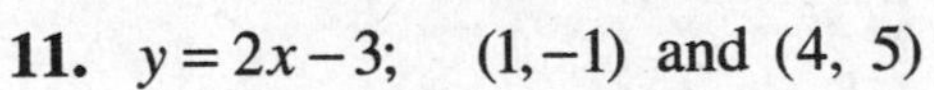

11. $y = 2x - 3$; $(1, -1)$ and $(4, 5)$

11. ______________

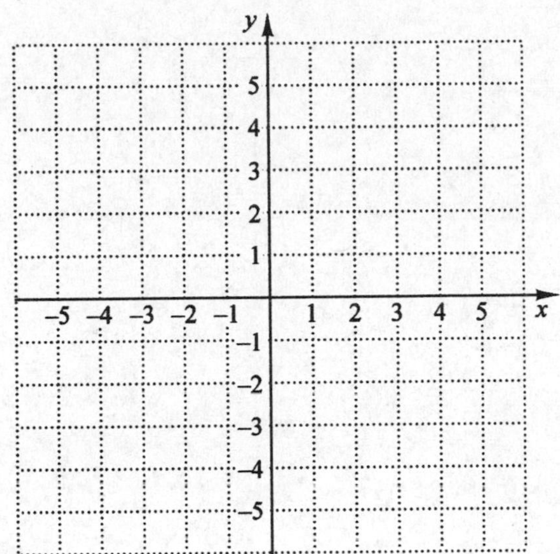

12. $x + 3y = 5$; $(2, 1)$ and $(5, 0)$

12. ______________

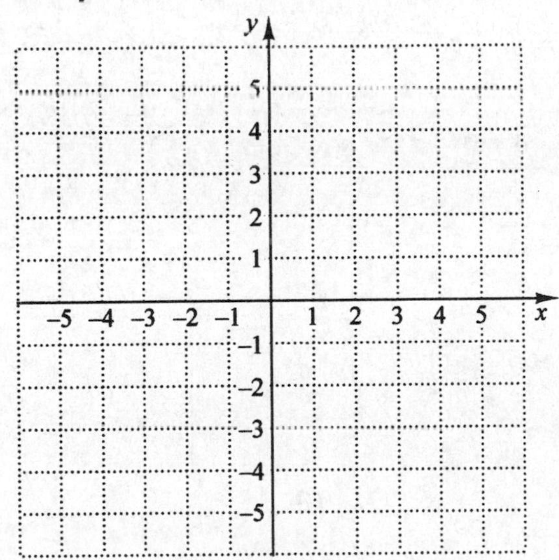

Objective c Graph linear equations using tables.

Graph.

13. $y = \frac{2}{3}x$

x	y

14. $y = x - 2$

x	y

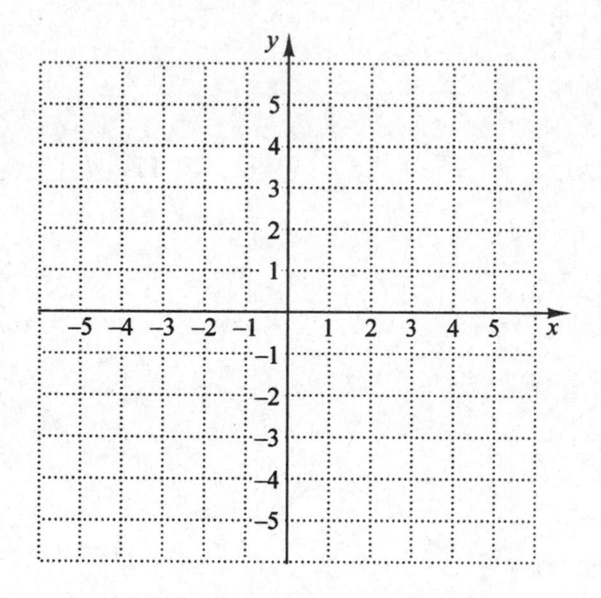

Name: Date:
Instructor: Section:

15. $x + y = 3$

x	y

16. $x + 4y = 4$

x	y

17. $5x - 2y = 10$

x	y

Objective d Graph nonlinear equations using tables.

Graph.

18. $y = |x| - 1$

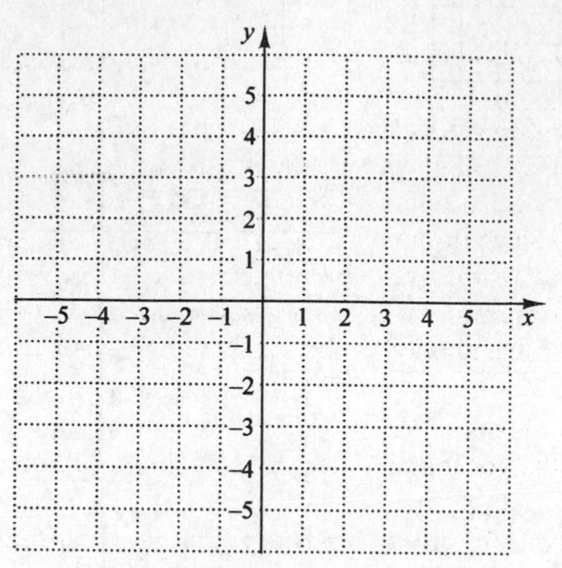

19. $y = x^2 - 4$

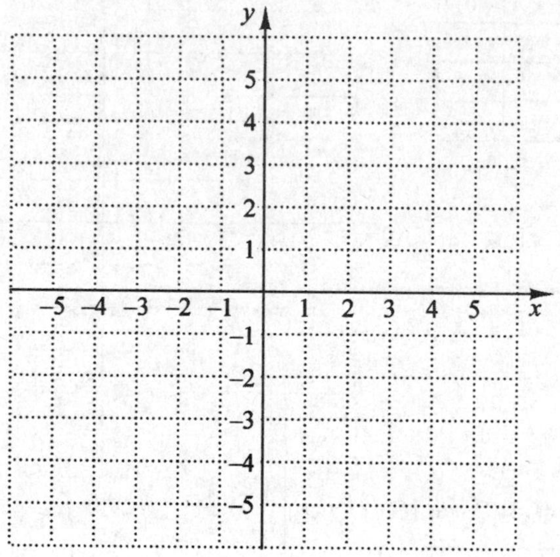

Name: Date:
Instructor: Section:

Chapter 2 GRAPHS, FUNCTIONS AND APPLICATIONS

2.2 Functions and Graphs

Learning Objective

a Determine whether a correspondence is a function.
b Given a function described by an equation, find function values (outputs) for specified values (inputs).
c Draw the graph of a function.
d Determine whether a graph is that of a function using the vertical-line test.
e Solve applied problems involving functions and their graphs.

Key Terms

Use the vocabulary terms listed below to complete each statement in Exercises 1–5.

function **input** **independent** **relation** **outputs**

1. A(n) ____________________ is a member of the domain of a function.

2. A ____________________ is a rule that assigns to each member of some set exactly one member of another set.

3. Function values are also called ____________________.

4. The variable x in $f(x) = 2x - 7$ is called the ____________________ variable.

5. A ____________________ is a rule that assigns to each member of some set at least one member of another set.

Objective a Determine whether a correspondence is a function.

Determine whether each correspondence is a function.

6.

−5 → −4
−3 → 0
−1 → 1
2 → 5
3 → 5

6. ______________

7.

−5 → −4
−3 → 0
−1 → 3
−1 → 5
2 → 5

7. ______________

8. Determine if the relation is a function.
Domain: Each person in a town
Correspondence: Each person's uncle
Range: A set of males

8. ______________

Objective b Given a function described by an equation, find function values (outputs) for specified values (inputs).

Find the function values.

9. $g(v) = 7v + 1$; find $g(4)$, $g(-2)$, and $g(6.5)$.

9. ______________

10. $g(x) = -2x - 7$; find $g(-2)$.

10. ______________

11. $f(x) = 3x^2 - 3x$; find $f(0)$, $f(-1)$, and $f(2)$.

11. ______________

12. As the price of a product increases, the consumer's purchases, or demand, for the product decreases. Suppose that under certain conditions in our economy, the demand for sugar is related to price by the demand function $d(p) = -1.6p + 24.1$, where p is the price of a 5-lb bag of sugar and $d(p)$ is the quantity of 5-lb bags, in millions, purchased at price p. What is the quantity purchased when the price is \$1 per 5-lb bag? \$2 per 5-lb bag?

12. ______________

Name: Date:
Instructor: Section:

Objective c Draw the graph of a function.

Graph each function.

13. $f(x) = 3x + 1$

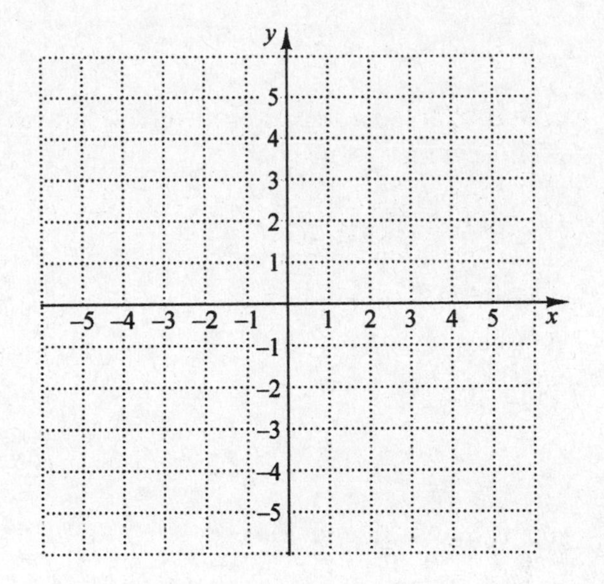

14. $f(x) = -\frac{2}{3}x + 3$

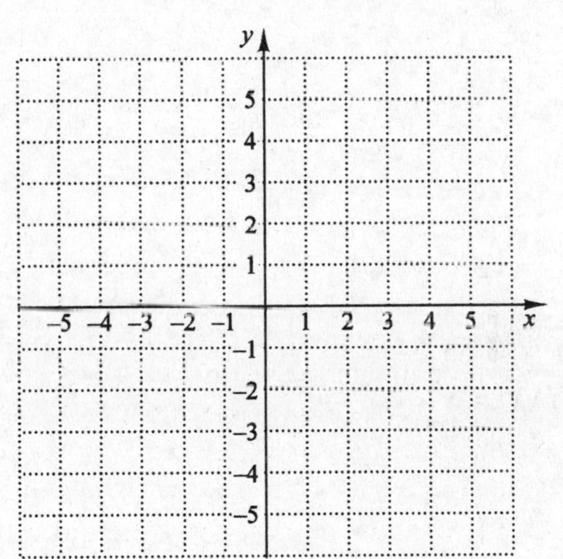

15. $g(x) = 4|x|$

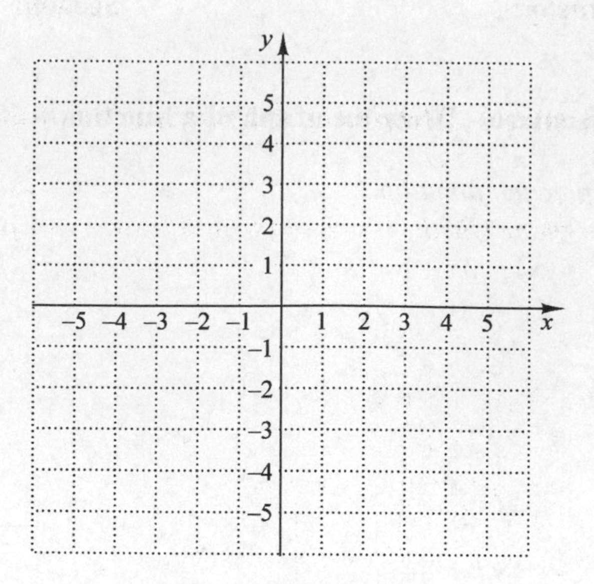

16. $f(x) = 3 - x^2$

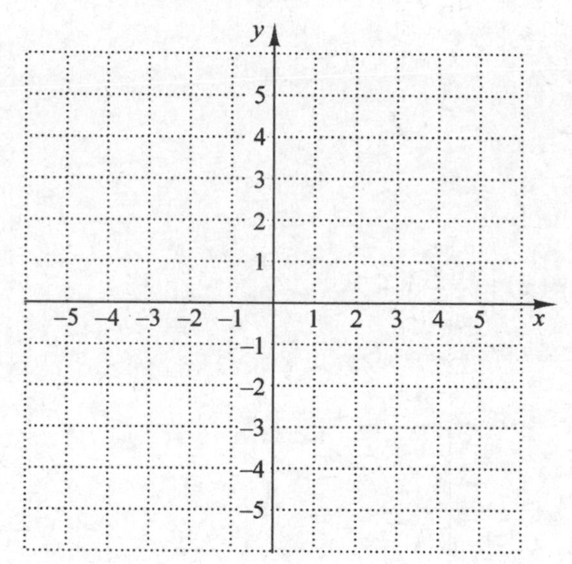

17. $f(x) = x^2 - 4x + 4$

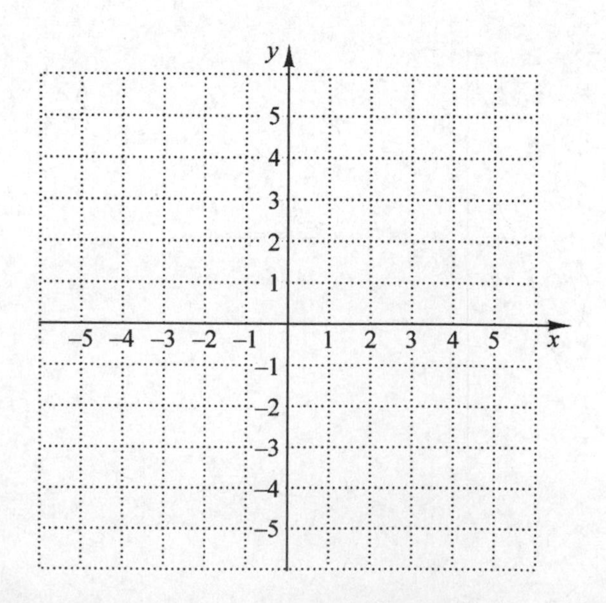

Name: Date:
Instructor: Section:

Objective d Determine whether a graph is that of a function using the vertical-line test.

Determine whether each of the following is the graph of a function.

18.

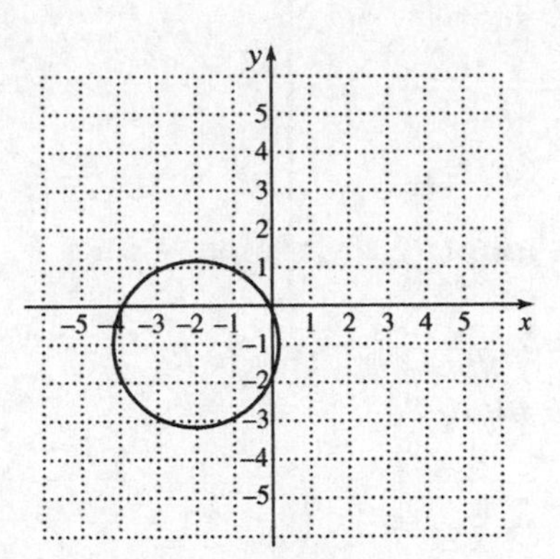

18. ______________

19.

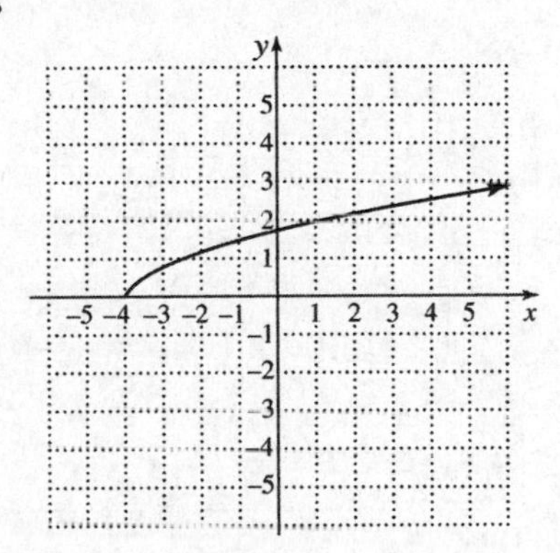

19. ______________

20.

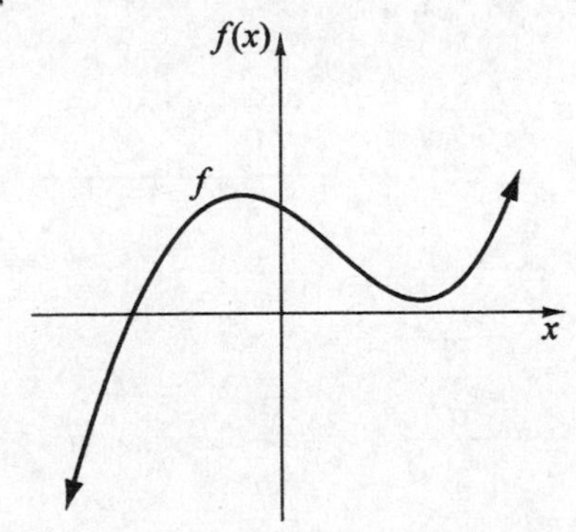

20. ______________

21.

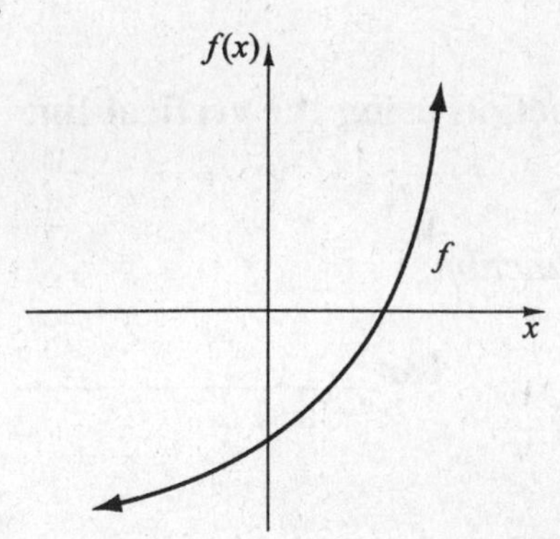

21. ______________

Objective e Solve applied problems involving functions and their graphs.

The following graph approximates the number of words participants were able to memorize. The number of words is a function f of the time, in minutes.

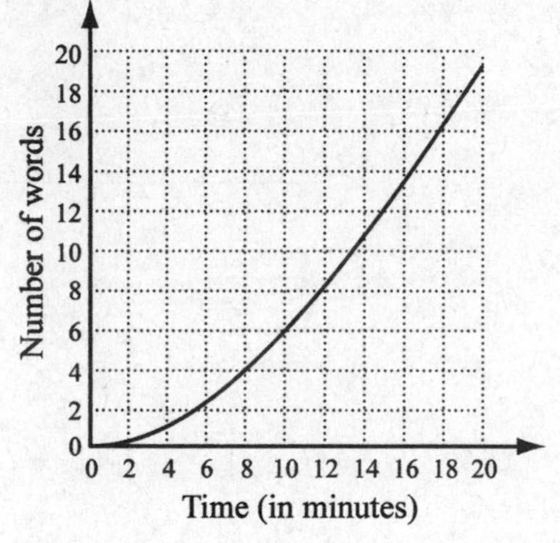

22. Approximate the number of words memorized after 10 minutes. That is, find $f(10)$.

22. ______________

23. Approximate the number of words memorized after 18 minutes. That is, find $f(18)$.

23. ______________

Name: Date:
Instructor: Section:

Chapter 2 GRAPHS, FUNCTIONS AND APPLICATIONS

2.3 Finding Domain and Range

Learning Objectives
a Find the domain and the range of a function.

Key Terms
Use the vocabulary terms listed below to complete each statement in Exercises 1–4.

domain **function** **relation** **range**

1. The __________________ is the set of all first coordinates.

2. The __________________ is the set of all second coordinates.

3. A set of ordered pairs is called a __________________ .

4. When a set of ordered pairs is such that no two different pairs share a common first coordinate, we have a __________________ .

Objective a Find the domain and the range of a function.

In Exercises 5-7, the graph is that of a function. Determine for each one **(a)** $f(1)$; **(b)** *the domain;* **(c)** *any x-values for which* $f(x)=2$; *and* **(d)** *the range.*

5.

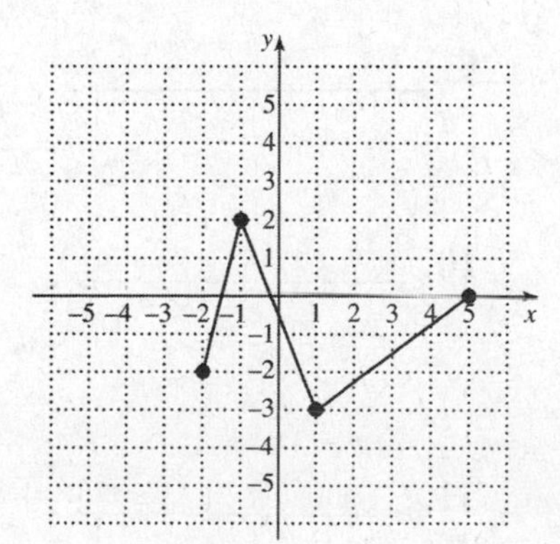

5.
a.__________________

b.__________________

c.__________________

d.__________________

6.

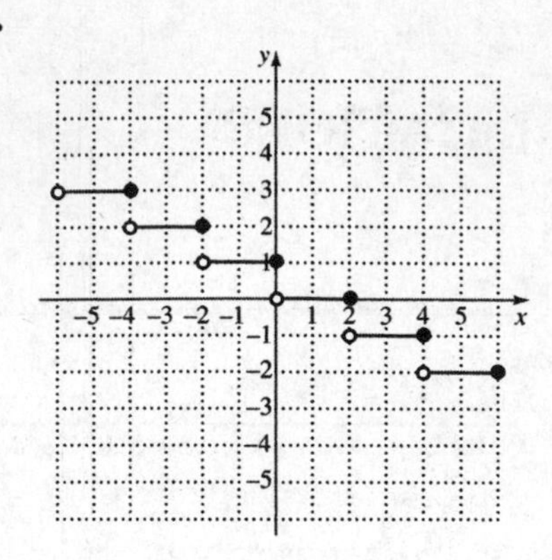

6.

a.____________________

b.____________________

c.____________________

d.____________________

7.

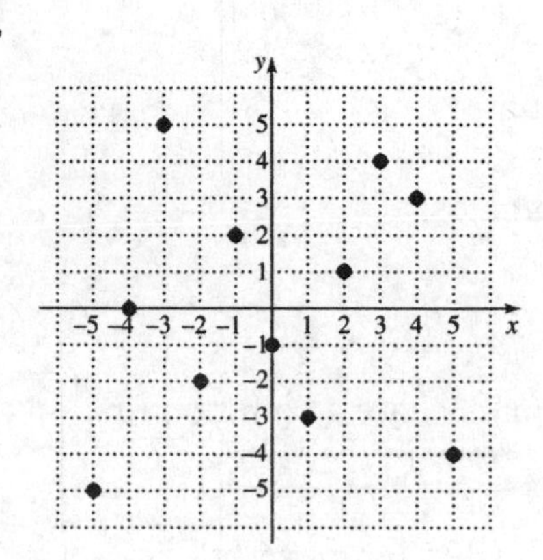

7.

a.____________________

b.____________________

c.____________________

d.____________________

Find the domain.

8. $f(x)=\dfrac{7}{x-1}$

8. ____________________

9. $f(x)=|5x+4|$

9. ____________________

10. $g(x)=\dfrac{4}{5}x$

10. ____________________

11. $g(x)=\dfrac{3}{|7x-5|}$

11. ____________________

Name: Date:
Instructor: Section:

Chapter 2 GRAPHS, FUNCTIONS AND APPLICATIONS

2.4 Linear Functions: Graphs and Slope

Learning Objectives

a Find the y-intercept of a line from the equation $y = mx + b$ or $f(x) = mx + b$.

b Given two points on a line, find the slope. Given a linear equation, derive the equivalent slope-intercept equation and determine the slope and the y-intercept.

c Solve applied problems involving slope.

Key Terms

Use the vocabulary terms listed below to complete each statement in Exercises 1–7.

run **rise** **grade** **up** **down** ***m*** ***b***

1. The slope of any line written in the form $y = mx + b$ is ________________.

2. The y-intercept of any line written in the form $y = mx + b$ is ________________.

3. Lines with positive slope slant ________________ from left to right.

4. Lines with negative slope slant ________________ from left to right.

5. The change in y as we move from one point to another is called the ________________.

6. The change in x as we move from one point to another is called the ________________.

7. The ________________ of the road is a measure of how steep it is.

Objective a Find the y-intercept of a line from the equation $y = mx + b$ or $f(x) = mx + b$.

Objective b Given two points on a line, find the slope. Given a linear equation, derive the equivalent slope-intercept equation and determine the slope and the y-intercept.

Find the slope and the y-intercept.

8. $y = -\frac{3}{5}x + 2$ **8.** ________________

9. $4x - 5y = 9$ **9.** ________________

10. $y = \frac{4}{9}x - 5$

10. ______________

11. $2x + 3y = 9$

11. ______________

12. $g(x) = -3.5x$

12. ______________

13. $5x - 3y = -9$

13. ______________

Objective b Given two points on a line, find the slope. Given a linear equation, derive the equivalent slope-intercept equation and determine the slope and the y-intercept.

Find the slope of each line.

14.

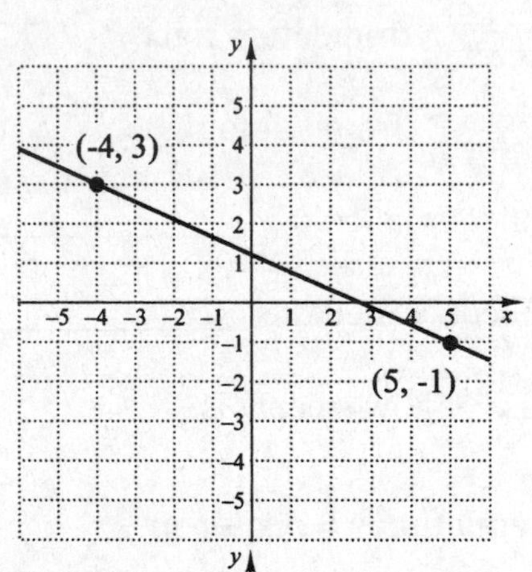

14. ______________

15.

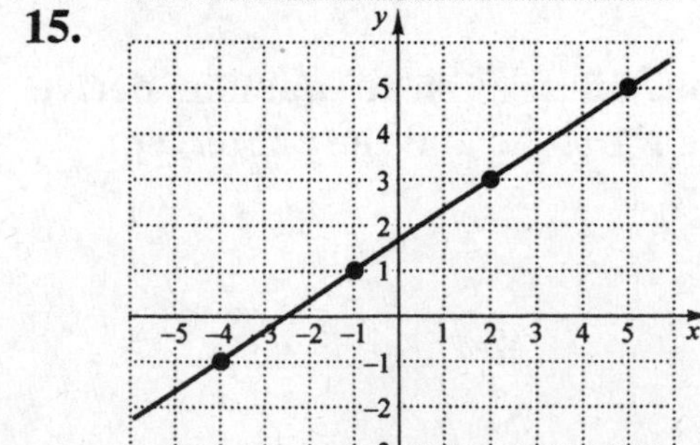

15. 

Name: Date:
Instructor: Section:

Find the slope of the line containing the given pair of points.

16. $\left(3,-\frac{1}{3}\right),\ \left(-2,\ \frac{5}{3}\right)$ **16.** ______________

17. $\left(\frac{1}{2},-2\right),\ \left(-\frac{1}{2},-5\right)$ **17.** ______________

18. $(0.6,-0.8),\ (0.1,-0.8)$ **18.** ______________

19. $(10,\ 15),\ (-25,-8)$ **19.** ______________

Objective c Solve applied problems involving slope.

Find the slope (or rate of change).

20. A road rises 7 ft over a horizontal distance of 280 ft. Find the slope (or grade) of the road. **20.** ______________

21. A ramp rises 3 ft over a horizontal distance of 50 ft. Find the grade of the ramp. **21.** ______________

Find the rate of change.

22.

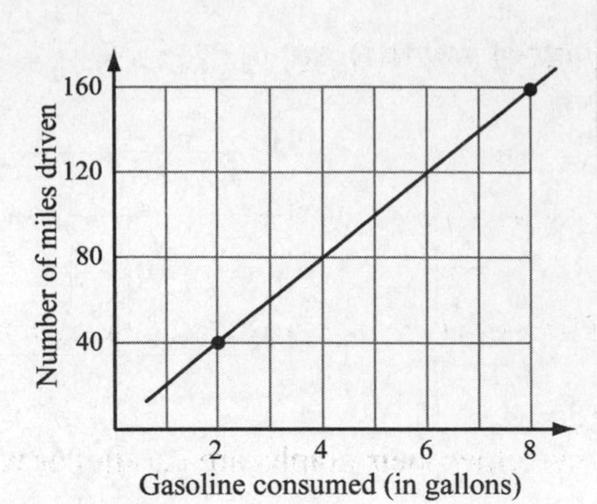

22. ______________

23.

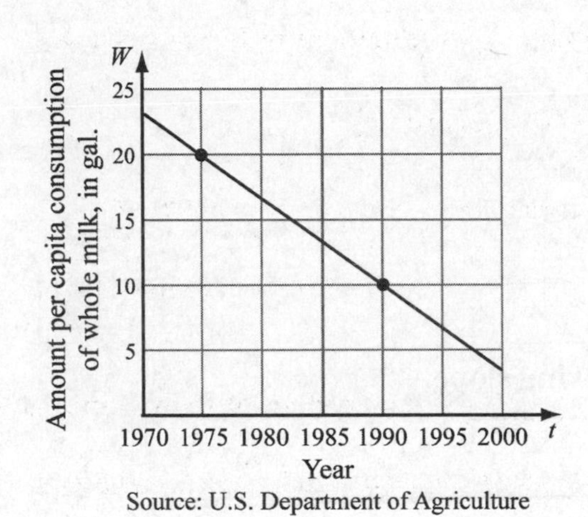

23. ______________

24.

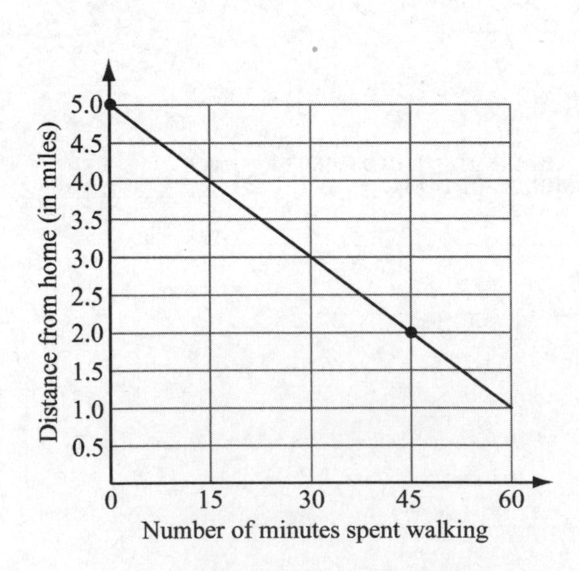

24. ______________

Name: Date:
Instructor: Section:

Chapter 2 GRAPHS, FUNCTIONS AND APPLICATIONS

2.5 More on Graphing Linear Equations

Learning Objective

a Graph linear equations using intercepts.
b Given a linear equation in slope-intercept form, use the slope and the y-intercept to graph the line.
c Graph linear equations of the form $x = a$ or $y = b$.
d Given the equations of two lines, determine whether their graphs are parallel or whether they are perpendicular.

Key Terms

Use the vocabulary terms listed below to complete each statement in Exercises 1–8.

not defined	***x*-intercept**	***y*-intercept**	**0**
horizontal	**parallel**	**perpendicular**	**vertical**

1. Two nonvertical lines are ____________________ if they have the same slope and different y-intercepts.

2. Two nonvertical lines are ____________________ if the product of their slopes is -1.

3. Parallel ____________________ lines have equations $x = p$ and $x = q$, where $p \neq q$.

4. If one equation in a pair of perpendicular lines is vertical, then the other is ____________________.

5. The ____________________ occurs where a graph crosses the x-axis.

6. The ____________________ occurs where a graph crosses the y-axis.

7. The slope of a horizontal line is ________________________.

8. The slope of a vertical line is ________________________.

Objective a Graph linear equations using intercepts.

Find the intercepts and then graph the line.

9. $x + 2y = 4$

9. ______________

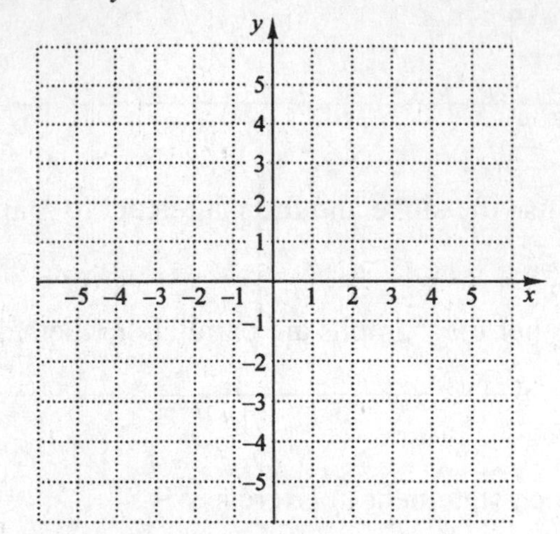

10. $3y - 3 = x$

10. ______________

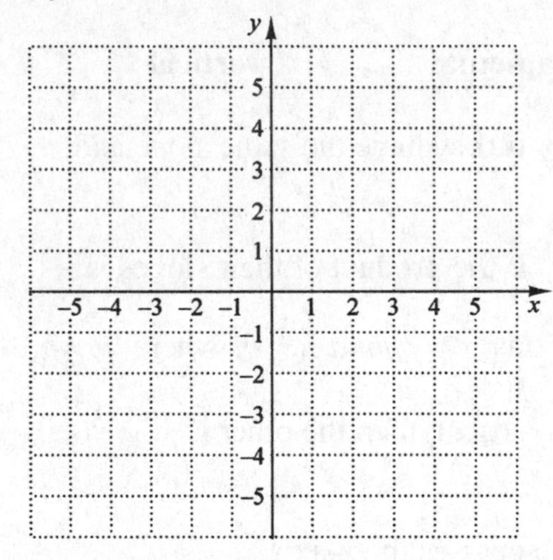

11. $x + 5 = y$

11. ______________

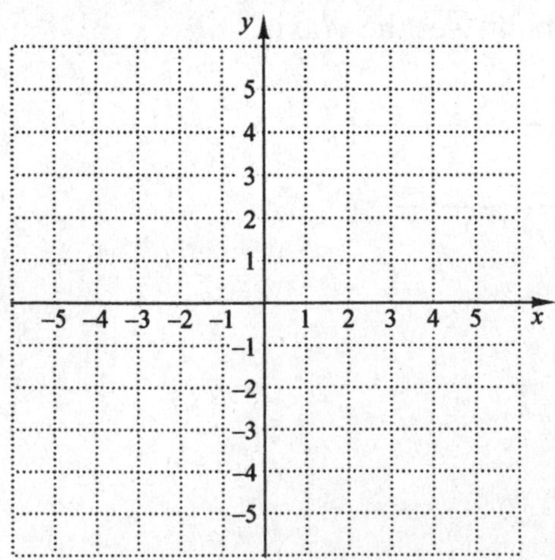

Name: Date:
Instructor: Section:

12. $4x - 5y = 20$

12. ________________

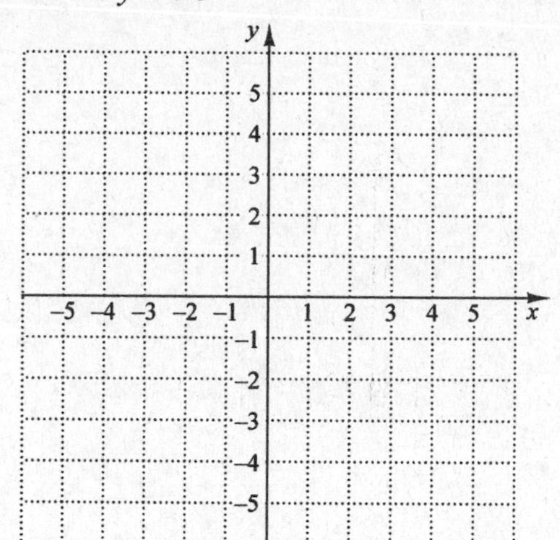

Objective b Given a linear equation in slope-intercept form, use the slope and the *y*-intercept to graph the line.

Graph using the slope and the y-intercept.

13. $4x + 2y = 8$

13. ________________

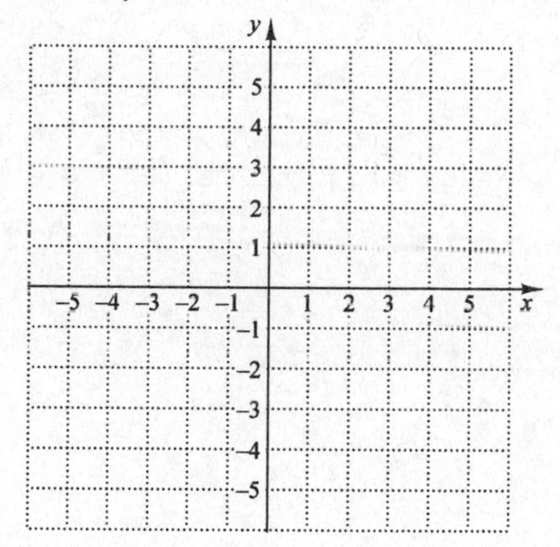

14. $y = -4 - 4x$

14. ______________

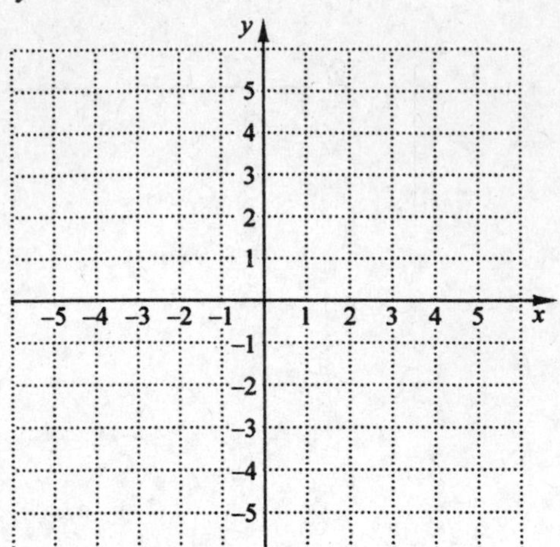

Objective c Graph linear equations of the form $x = a$ or $y = b$.

Graph and, if possible, determine the slope.

15. $x = -3$

15. ______________

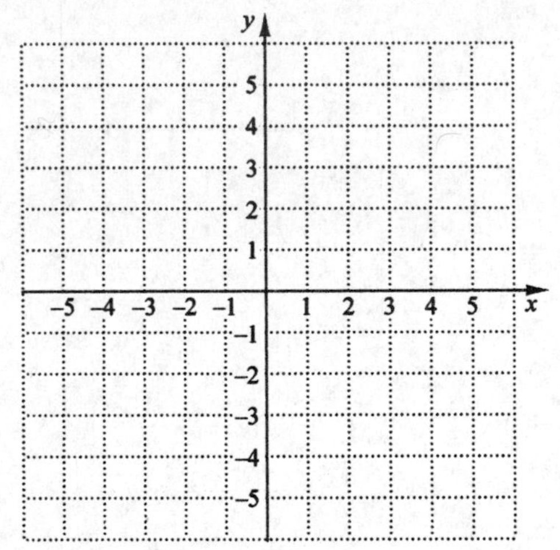

Name: Date:
Instructor: Section:

16. $y = 4$

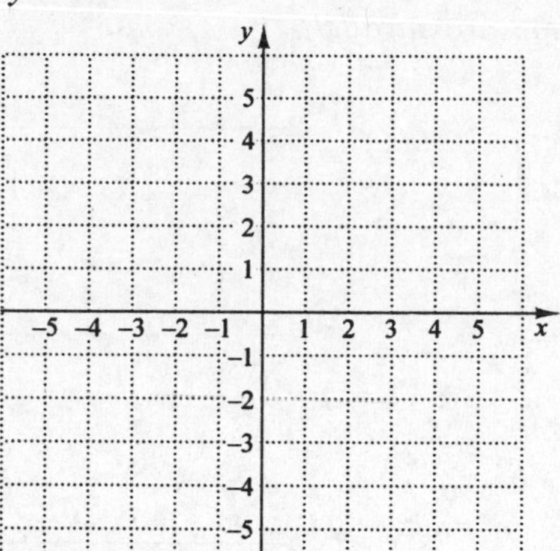

16. ______________

17. $2y = -5$

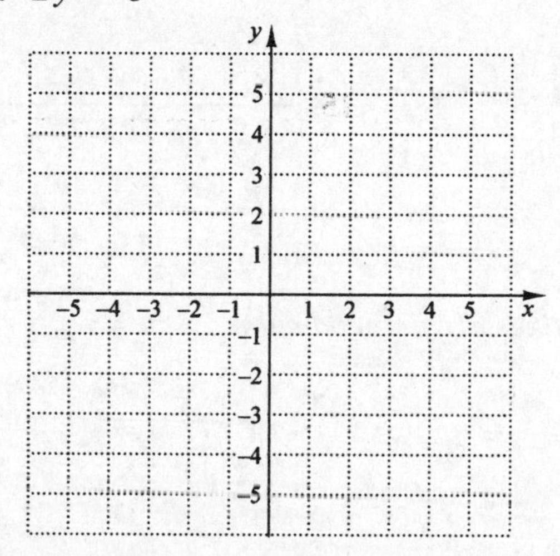

17. ______________

18. $3x - 15 = 0$

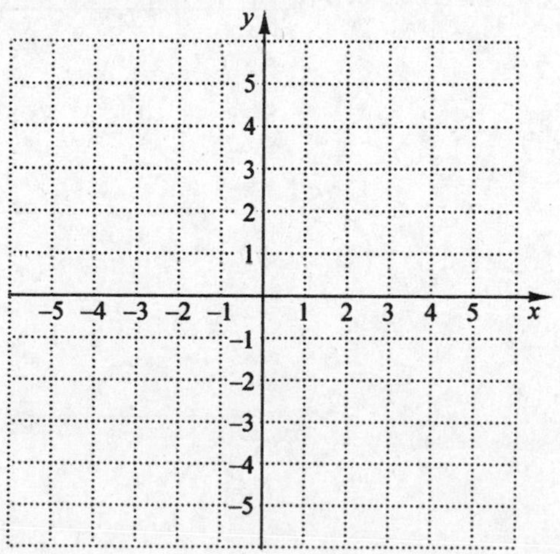

18. ______________

Objective d Given the equations of two lines, determine whether their graphs are parallel or whether they are perpendicular.

Determine whether the graphs of the given pair of lines are parallel.

19. $x - y = 7,$
$y = x + 3$

19. ______________

20. $y = 2x - 3,$
$4x - 2y = 1$

20. ______________

21. $y + 2 = 5x,$
$5x + y = 3$

21. ______________

22. $y = 3,$
$y = -5$

22. ______________

Determine whether the graphs of the given pair of lines are perpendicular.

23. $y = 5 - 2x,$
$2y + x = 1$

23. ______________

24. $x - y = 3,$
$x = y + 5$

24. ______________

25. $y = x + 1,$
$x + y = 4$

25. ______________

26. $x = 7,$
$y = 0$

26. ______________

Name: Date:
Instructor: Section:

Chapter 2 GRAPHS, FUNCTIONS AND APPLICATIONS

2.6 Finding Equations of Lines; Applications

Learning Objective

a Find an equation of a line when the slope and the y-intercept are given.
b Find an equation of a line when the slope and a point are given.
c Find an equation of a line when two points are given.
d Given a line and a point not on the given line, find an equation of the line parallel to the line and containing the point, and find an equation of the line perpendicular to the line and containing the point.
e Solve applied problems involving linear functions.

Key Terms

Use the vocabulary terms listed below to complete each statement in Exercises 1–6.

slope	**point-slope**	**slope-intercept**
***y*-intercept**	**parallel**	**perpendicular**

1. The ________________ form of a line is written $y - y_1 = m(x - x_1)$.

2. Two lines with different y-intercepts are ________________ if they have the same slope.

3. Two lines are ________________ if the product of their slopes is -1.

4. The equation $y = mx + b$ is in ________________ form.

5. The graph of an equation $y = mx + b$ has ________________ m.

6. The graph of an equation $y = mx + b$ has ________________ $(0, b)$.

Objective a Find an equation of a line when the slope and the y-intercept are given.

Find an equation of the line having the given slope and y-intercept.

7. Slope $= -15$, y-intercept $= (0, 12)$ 7. ________________

8. Slope $= 2.7$, y-intercept $= (0, -4.3)$ 8. ________________

Find a linear function $f(x) = mx+b$ whose graph has the given slope and y-intercept.

9. Slope $= -\frac{5}{6}$, *y*-intercept $= (0,-8)$ **9.** ____________

10. Slope $= \frac{3}{4}$, *y*-intercept $= \left(0, -\frac{5}{11}\right)$ **10.** ____________

Objective b Find an equation of a line when the slope and a point are given.

Find an equation of the line having the given slope and containing the given point.

11. $(-4, 0)$, $m = -3$ **11.** ____________

12. $(1, 6)$, $m = \frac{2}{3}$ **12.** ____________

13. $(5, -1)$, $m = 2$ **13.** ____________

14. $(0, 7)$, $m = -5$ **14.** ____________

Objective c Find an equation of a line when two points are given.

Find an equation of the line containing the given pair of points.

15. $(3, 8)$ and $(1, 10)$ **15.** ____________

16. $(0, 3)$ and $(5, 2)$ **16.** ____________

Name: Date:
Instructor: Section:

17. $(4, 7)$ and $(-1, 3)$ **17.** ________________

18. $(-7, 5)$ and $(-1, 3)$ **18.** ________________

Objective d Given a line and a point not on the given line, find an equation of the line parallel to the line and containing the point, and find an equation of the line perpendicular to the line and containing the point.

Write an equation of the line containing the given point and parallel to the given line.

19. $(2,-1)$; $4x-y=8$ **19.** ________________

20. $(-12,0)$; $2x+3y=5$ **20.** ________________

21. $(0,4)$; $y=8x-9$ **21.** ________________

Write an equation of the line containing the given point and perpendicular to the given line.

22. $(-3,6)$; $5x+10y=2$ **22.** ________________

23. $(0,8)$; $6x+7y=5$ **23.** ________________

24. $(0,3)$; $y=x-7$ **24.** ________________

Objective e Solve applied problems involving linear functions.

Solve.

25. In 1992, 1250 students attended Eastside College. By 2007, the college had 1425 students. Let $c(t)$ represent the number of students at the college t years after 1992.

a. Find a linear function that fits the data.

b. Use the function of part (a) to predict the number of students in 2010.

c. When will 1600 students attend Eastside College?

25.
a.______________
b.______________
c.______________

26. In 1985, the record for the 100-m run at Eastside College was 12.3 sec. In 2005, it was 12.1 sec. Let $r(t)$ represent the record in the 100-m run and t the number of years since 1985.

a. Find a linear function that fits the data.

b. Use the function of part (a) to predict the record in 2000 and in 2010.

c. Find the year when the record will be 12.0 sec.

26.
a.______________
b.______________
c.______________

Name: Date:
Instructor: Section:

Chapter 3 SYSTEMS OF EQUATIONS

3.1 Systems of Equations in Two Variables

Learning Objectives

a Solve a system of two linear equations or two functions by graphing and determine whether a system is consistent or inconsistent and whether the equations in a system are dependent or independent.

Key Terms

Use the vocabulary terms listed below to complete each statement in Exercises 1–6.

consistent	**dependent**	**inconsistent**
independent	**intersection**	**system**

1. A(n) ____________________ of equations is a set of two or more equations, in two or more variables, for which a common solution is sought.

2. When solving a system of equations graphically, we look for the ____________________ of the graphs of the equations.

3. A system of equations that has at least one solution is said to be ____________________.

4. A system of equations with no solution is called ____________________.

5. If a system of two equations in two variables has infinitely many solutions, the equations are ____________________.

6. If a system of two equations in two variables has one solution or no solutions, the equations are ____________________.

Objective a **Solve a system of two linear equations or two functions by graphing and determine whether a system is consistent or inconsistent and whether the equations in a system are dependent or independent.**

Solve each system of equations graphically. Then classify the system as consistent or inconsistent and the equations as dependent or independent.

7. $x - y = 1,$
$x + y = 5$

7. 

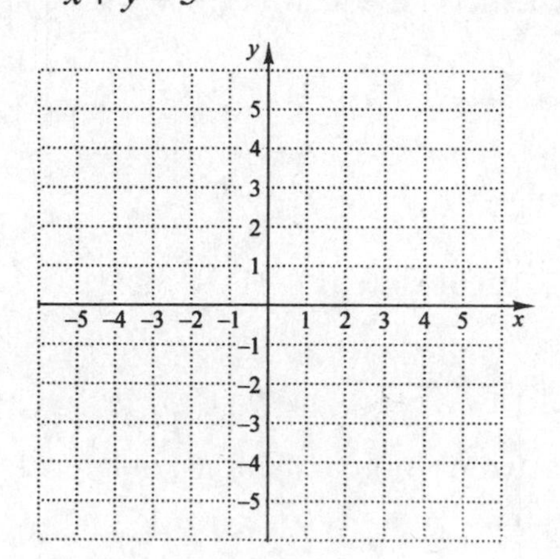

8. $6x - y = 1,$
$3x + y = 8$

8. 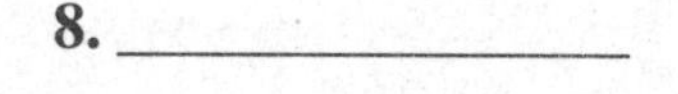

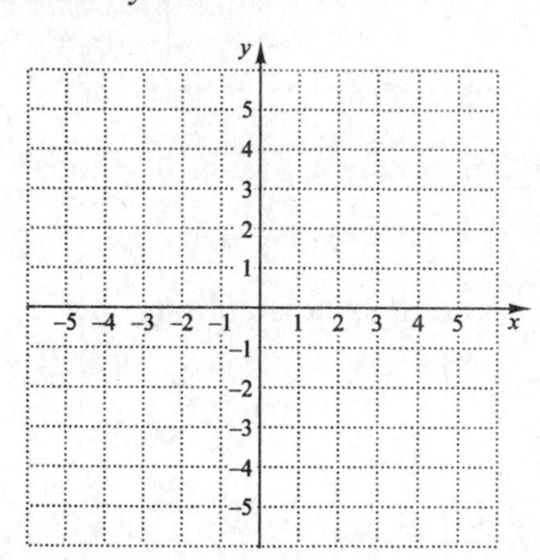

Name: Date:
Instructor: Section:

9. $y = \frac{1}{2}x,$
$x = 3y + 1$

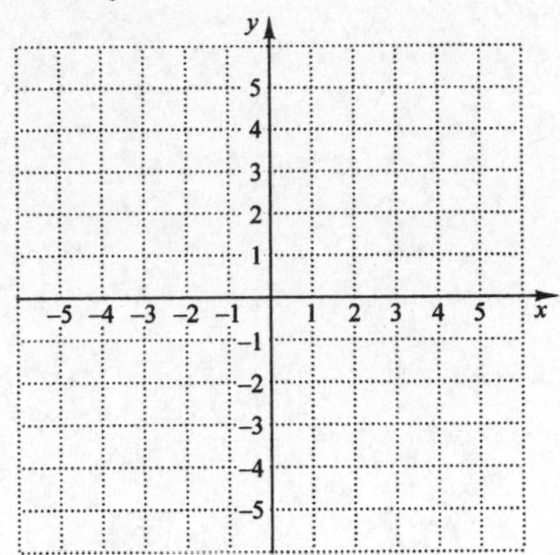

9. ________________

10. $x = 2y,$
$x - 2y = 2$

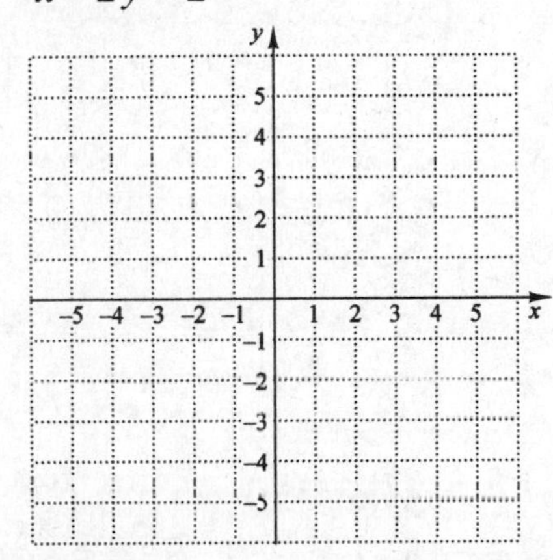

10. ________________

11. $x + y = 3,$
$x = 3 - y$

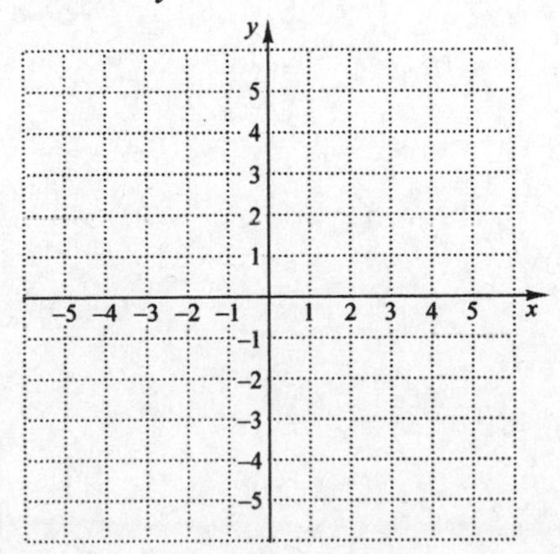

11. ________________

12. $x - \frac{1}{2}y = 3,$

$x = 3y - 2$

12. ________________

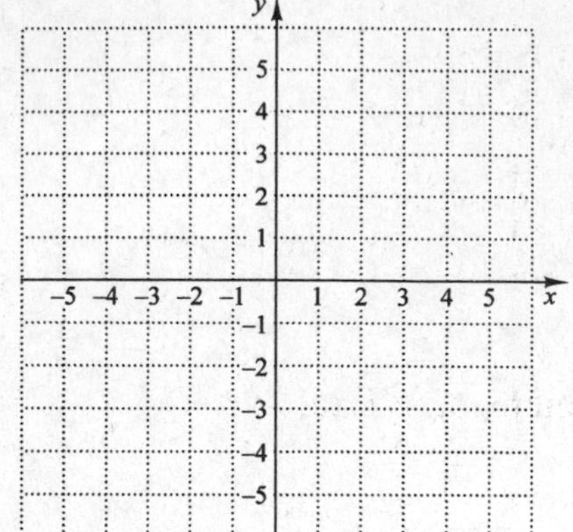

13. $2x - y = 2,$

$4x = 1 - y$

13. 

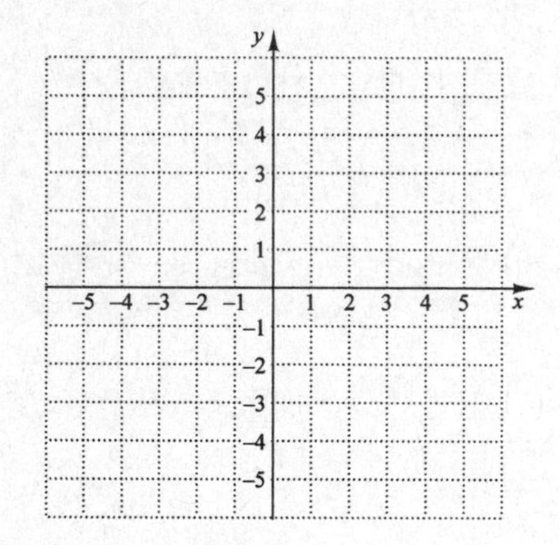

14. 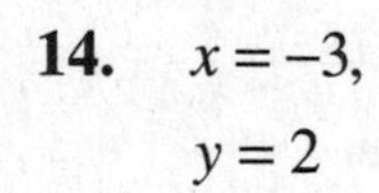

$x = -3,$

$y = 2$

14. 

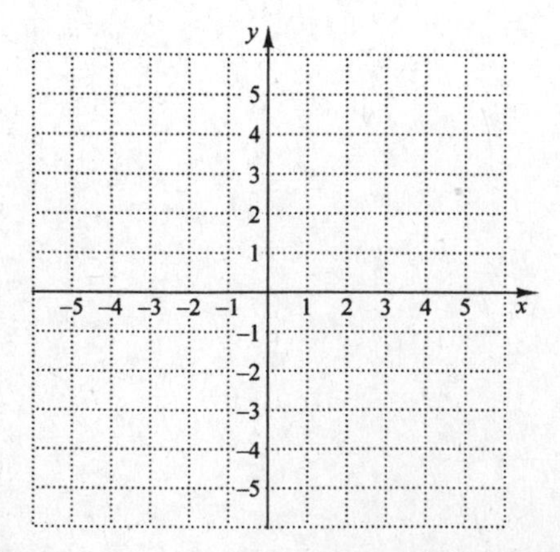

Name: Date:
Instructor: Section:

Chapter 3 SYSTEMS OF EQUATIONS

3.2 Solving by Substitution

Learning Objective

a Solve systems of equations in two variables by the substitution method.

b Solve applied problems by solving systems of two equations using substitution.

Key Terms

Use the vocabulary terms listed below to complete each statement in Exercises 1–4.

algebraic **solve** **substitute** **translate**

1. Nongraphical methods for solving systems of equations are called ________________ methods.

2. One way to solve a system of two equations is to ________________ an equivalent expression for a variable into an equation.

3. When using the substitution method, if neither equation of a pair has a variable alone on one side, we ________________ one equation for one of the variables.

4. Sometimes it is easier to ________________ a problem into two equations in two variables.

Objective a Solve systems of equations in two variables by the substitution method.

Solve each system of equations by the substitution method.

5. $x+y=5$,
$y=x+1$

5.________________

6. $x=y+6$,
$x+2y=12$

6.________________

7. $y = 3x - 2$,
$2y - x = 1$

7.________________

8. $2x + 5y = 9$,
$x = y - 1$

8.________________

9. $a + b = -6$,
$b - a = 8$

9.________________

10. $3x - y = 3$,
$-5x + y = 3$

10.________________

11. $s - 6t = 5$,
$3s + 2t = 0$

11.________________

Name: Date:
Instructor: Section:

12. $3y+4=x$,
$3y-x=5$

12.________________

13. $p-q=-5$,
$2p=q+1$

13.________________

14. $x-3y=7$,
$3x-5y=1$

14.________________

15. $7x+2y=-4$,
$8x-y=2$

15.________________

Objective b Solve applied problems by solving systems of two equations using substitution.

Solve.

16. The perimeter of the largest possible regulation-size rugby field is 428 m. The length is 4 m longer than twice the width. Find the dimensions.

16. ______________

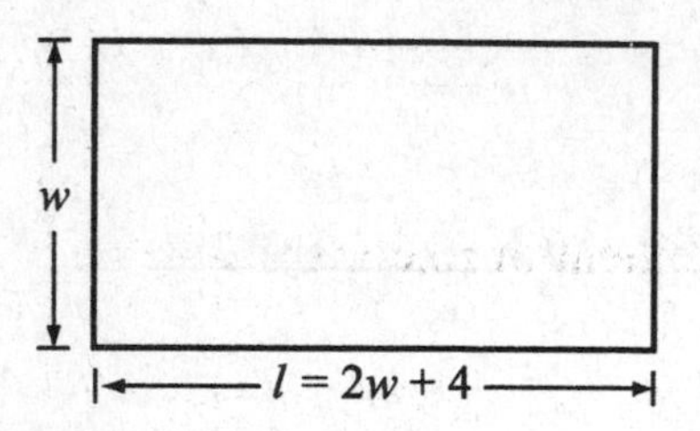

17. A rectangle has a perimeter of 124 ft. The width is 10 ft less than the length. Find the length and width.

17. 

18. The perimeter of a doubles-play badminton court is 128 ft. The width is 2 ft less than half the length. Find the length and the width.

18. ______________

Name: Date:
Instructor: Section:

Chapter 3 SYSTEMS OF EQUATIONS

3.3 Solving by Elimination

Learning Objectives
a Solve systems of equations in two variables by the elimination method.
b Solve applied problems by solving systems of two equations using elimination.

Key Terms
Use the vocabulary terms listed below to complete each statement in Exercises 1–4.

elimination **false** **opposites** **true**

1. The ________________ method for solving systems of equations makes use of the addition principle.

2. To eliminate a variable when adding, the terms containing that variable must be ________________.

3. When solving using the elimination method, obtaining a(n) ________________ equation means there is no solution.

4. When solving using the elimination method, obtaining a(n) ________________ equation means the system has an infinite number of solutions.

Objective a Solve systems of equations in two variables by the elimination method.

Solve each system of equations by the elimination method.

5. $x - y = 3,$
$x + y = 11$

5.________________

6. $a+b=8$,
$-a+5b=7$

6.________________

7. $3x-y=3$,
$-5x+y=3$

7.________________

8. $r+s=9$,
$3r-2s=-5$

8.________________

9. $8a+12b=10$,
$6a+9b=5$

9.________________

Name: Date:
Instructor: Section:

10. $3x-2y=5$,
$6x-10=4y$

10.________________

11. $6x-0.5y=3$,
$1.5x+2.25y=-4$

11.________________

12. $0.12s-0.06t=12$,
$0.08s+0.16t=-16$

12.________________

Objective b Solve applied problems by solving systems of two equations using elimination.

Solve. Use the elimination method when solving the translated system.

13. In a recent basketball game, the Westside Wildcats scored 91 of their points on a combination of 41 two- and three-point baskets. How many of each type of shot were made?

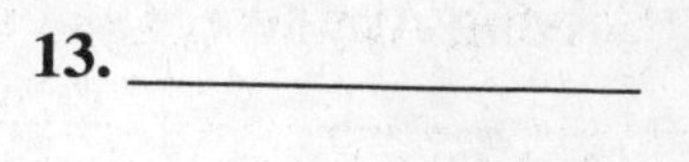

13. ________________

14. Supplementary angles are angles whose sum is $180°$. Two angles are supplementary. One angle measures $20°$ more than four times the measure of the other. Find the measure of each angle.

14. ________________

15. Complementary angles are angles whose sum is $90°$. Two angles are complementary. One angle is $10°$ less than four times the other. Find the measure of each angle.

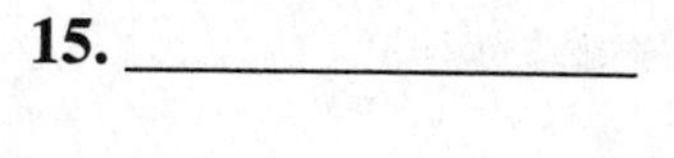

15. ________________

Name: Date:
Instructor: Section:

Chapter 3 SYSTEMS OF EQUATIONS

3.4 Solving Applied Problems: Two Equations

Learning Objectives

a Solve applied problems involving total value and mixture using systems of two equations.

b Solve applied problems involving motion using systems of two equations.

Key Terms

Use the vocabulary terms listed below to complete each numbered blank in the statement.

rate **motion** **distance** **time**

The ________________ formula states that ________________ =
(1) (2)

________________ (or speed) · ________________ .
(3) (4)

1. ________________

2. ________________

3. ________________

4. ________________

Objective a Solve applied problems involving total value and mixture using systems of two equations.

Solve.

5. Admission to Mammoth Cave is $12 for adults and $8 for youth (*Source:* National Park Service). One day, 575 people entered the cave, paying a total of $5600. How many adults and how many youth entered the cave?

5. ________________

6. Alpine Trail Mix is 40% nuts and Meadows Trail Mix is 25% nuts. How much of Alpine and how much of Meadows should be mixed to form a 10-lb batch of trail mix that is 32% nuts?

6. ______________

7. Firehouse Cafe charges $6.40 for one bowl of soup and one vegetable tortilla wrap, and $16.60 for two bowls of soup and three wraps. Determine the cost of one bowl of soup and one vegetable tortilla wrap.

7. ______________

8. The Clarkstown Volunteer Fire Department served 325 chicken and noodle dinners. A child's plate cost $4.50 and an adult's plate cost $6.00. A total of $1672.50 was collected. How many of each type of plate was served?

8. ______________

Name: Date:
Instructor: Section:

9. Tropical Punch is 18% fruit juice and Caribbean Spring is 24% fruit juice. How many liters of each should be mixed to get 18 L of a mixture that is 20% juice?

9. ______________

10. Tia received $7.55 in change in quarters and dimes. There were 38 coins in all. How many of each kind did she receive?

10. ______________

Objective b Solve applied problems involving motion using systems of two equations.

Solve.

11. Two private airplanes leave an airport at the same time, both flying due east. One travels at 150 mph and the other travels at 165 mph. In how many hours will they be 45 mi apart?

11. ______________

12. A moving van leaves a rest stop on an interstate highway and travels north at 50 mph. An hour later, a car leaves the rest stop and travels north at 70 mph. When will the car catch up to the van?

12. ______________

13. Gavin's boat took 2 hr to make a trip downstream with a 4-mph current. The return trip against the same current took 3 hr. Find the speed of the boat in still water.

13. ______________

14. It takes a passenger train 1 hr less time than it takes a freight train to make the trip from Coastal City to Valley Village. The passenger train averages 72 km/h, while the freight train averages 54 km/h. How far is it from Coastal City to Valley Village?

14. ______________

15. Two cars leave at the same time and travel directly toward each other from points 100 mi apart at rates of 50 mph and 70 mph. When will they meet?

15. ______________

Name: Date:
Instructor: Section:

Chapter 3 SYSTEMS OF EQUATIONS

3.5 Systems of Equations in Three Variables

Learning Objective
a Solve systems of three equations in three variables.

Objective a Solve systems of three equations in three variables.

Solve.

1. $x - y - 2z = 1,$
$x - 5y + 2z = 5,$
$2x - 3y - 4z = 2$

1. ________________

2. $a + b - c = 4,$
$2a - b - 3c = 1,$
$a + 2b + 3c = -1$

2. ________________

3. $x - 2y - z = 4,$
$2x + 3y + 2z = 11,$
$2x - y + 4z = 5$

3. ________________

4. $4x - y - z = -1,$
$2x + 3y + z = 2,$
$2x - y - z = -2$

4. ________________

5. $$\begin{aligned} p-q+3r&=2,\\ 2p-8q+3r&=6,\\ 10p+4q-9r&=0 \end{aligned}$$

5. ________________

6. $$\begin{aligned} 2a+c&=5,\\ 5a-2b&=0,\\ b+2c&=-5 \end{aligned}$$

6. ________________

7. $$\begin{aligned} 2x+y\phantom{{}+z}&=1,\\ 4x+3y+z&=7,\\ 2y+z&=2 \end{aligned}$$

7. ________________

8. $$\begin{aligned} x+2y-3z&=8,\\ 2x+y-2z&=11,\\ x+5y-8z&=15 \end{aligned}$$

8. ________________

Name: Date:
Instructor: Section:

Chapter 3 SYSTEMS OF EQUATIONS

3.6 Solving Applied Problems: Three Equations

Learning Objective
a Solve applied problems using systems of three equations.

Objective a Solve applied problems using systems of three equations.

Solve.

1. The sum of three numbers is 78. The third is 4 more than the second. The first is 2 less than twice the second. Find the numbers.

1. ______________

2. In triangle ABC, the measure of angle A is twice the measure of angle B. The measure of angle C is 4° less than that of angle B. Find the angle measures.

2. ______________

3. The number of calories in a 100-g serving of fruit varies by the type of fruit. A snack consisting of one serving each of grapes, melon, and pears contains 124 calories. A snack consisting of two servings of grapes and one serving of pears contains 160 calories. A snack consisting of 2 servings of melon, one serving of grapes, and two servings of pears contains 188 calories. How many calories are in a 100-g serving of each type of fruit?

3. ______________

4. Surresh is going to buy a computer. The basic model with a memory upgrade and a larger monitor costs $890. The basic model with only a memory upgrade costs $840. The price of the basic model with a larger monitor is $800. Find the price of the basic model, the price of a memory upgrade, and the added cost of a larger monitor.

4. ______________

Name: Date:
Instructor: Section:

5. Polar Treats sells one-dip ice cream cones for \$2.25, two-dip cones for \$3.15, and three-dip cones for \$3.85. On Saturday, they sold 200 cones for a total of \$599.20. Altogether, Polar Treats used 380 dips of ice cream. How many of each size cone were sold?

5. ______________

6. Ragheda has \$11,200 invested in three retirement accounts. Last year, the first grew by 8%, the second by 10%, and the third by 5%. Her total earnings were \$848. The earnings from the third investment were \$80 less than the earnings from the second investment. How much was invested in each account?

6. ______________

7. Willoughby County spent $7.6 million last year for schools, roads, and law enforcement. The amount spent on roads was $0.4 million more than the amount spent on law enforcement. The amount spent on roads was $0.2 million less than one fourth of what was spent on schools. How much did the county spend on schools, roads, and law enforcement?

7. ______________

8. In their final home game, the Wildcats scored a total of 113 points on a combination of 2-point field goals, 3-point field goals, and 1-point foul shots. Altogether, the Wildcats made 59 baskets and 22 more 2-pointers than foul shots. How many shots of each kind were made?

8. ______________

Name: Date:
Instructor: Section:

Chapter 3 SYSTEMS OF EQUATIONS

3.7 Systems of Inequalities in Two Variables

Learning Objectives
a Determine whether an ordered pair of numbers is a solution of an inequality in two variables.
b Graph linear inequalities in two variables.
c Graph systems of linear inequalities and find coordinates of any vertices.

Key Terms
Use the vocabulary terms listed below to complete each statement in Exercises 1–4.

half-plane **linear inequality** **test point** **vertices**

1. The sentence $2x-3\leq-5$ is an example of a(n) ________________.

2. The graph of a linear inequality is a(n) ________________.

3. To determine which side of the boundary to shade as the graph of the solution set of an inequality, select a(n) ________________ not on the line.

4. If a system of inequalities has a graph that consists of a polygon and its interior, the corners of the graph are called ________________.

Objective a Determine whether an ordered pair of numbers is a solution of an inequality in two variables.

Determine whether the ordered pair is a solution of the given inequality.

5. $(2,-5)$; $2x+4y\leq-5$ **5.** ______________

6. $(20,12)$; $2y-3x>1$ **6.** ______________

Objective b Graph linear inequalities in two variables.

Graph each inequality on a plane.

7. $y \leq \frac{1}{3}x$

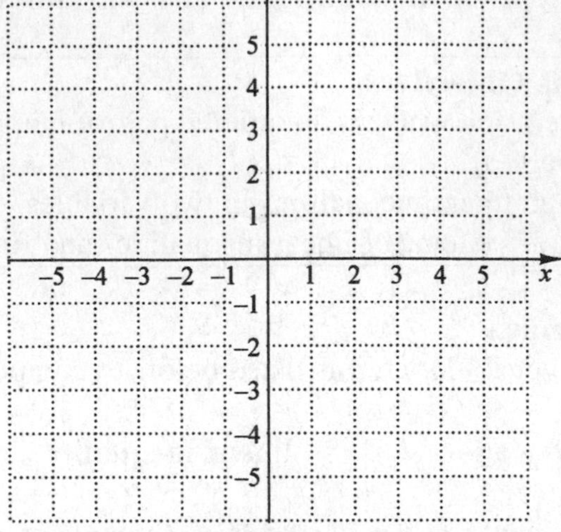

8. $y \geq x - 2$

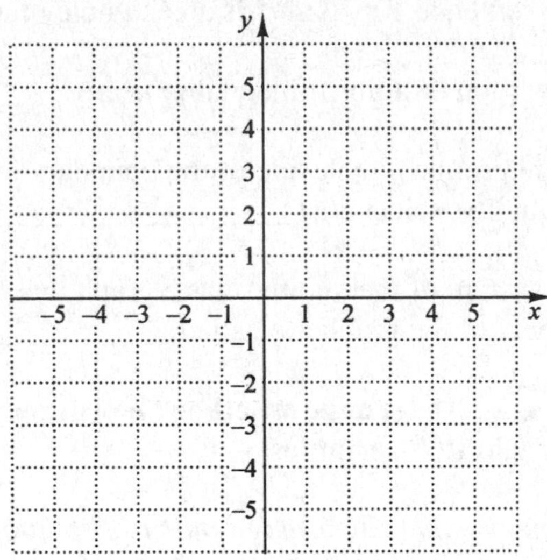

9. $4x - y > 8$

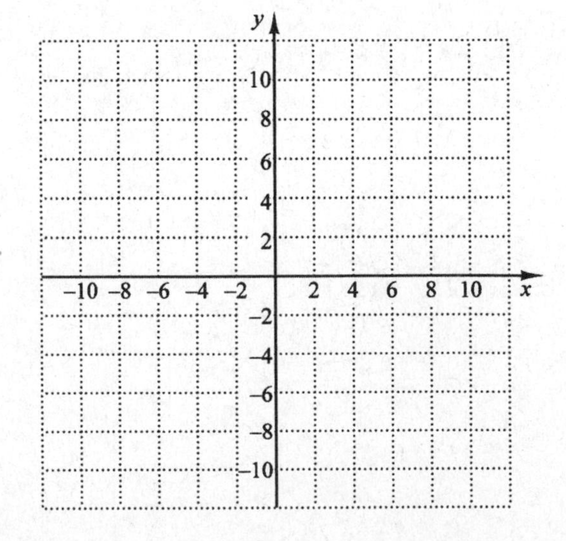

Name: Date:
Instructor: Section:

10. $x-3y<6-x$

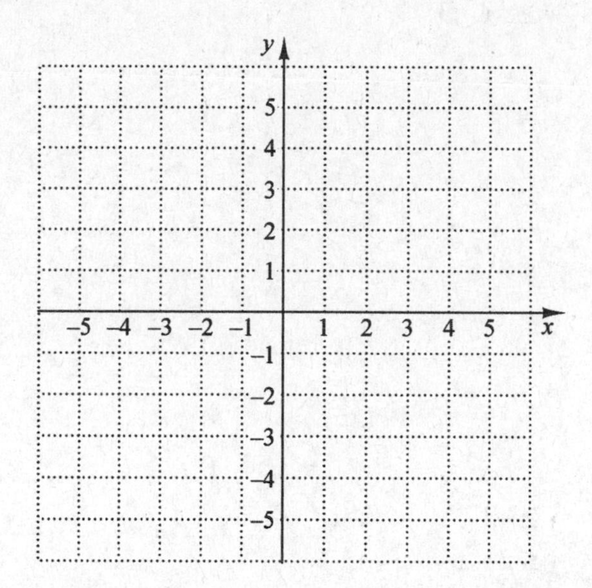

Objective c Graph systems of linear inequalities and find coordinates of any vertices.

Graph each system of inequalities.

11. $y<-x,$
$y>x-3$

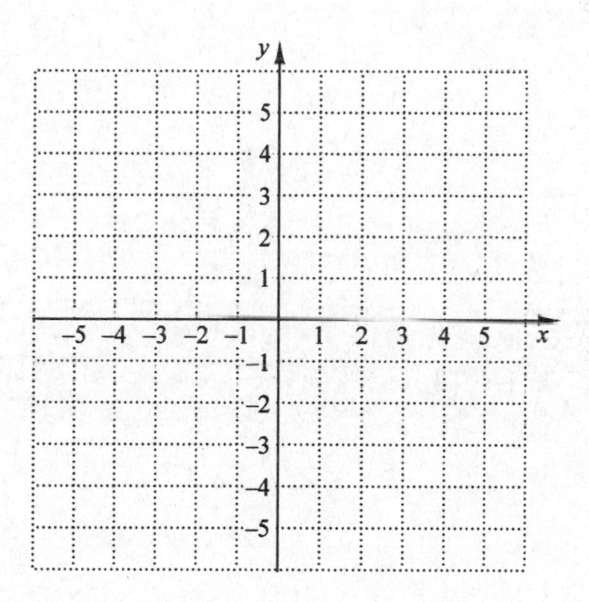

12. $y \le 2x$,
$y \ge 4x - 3$

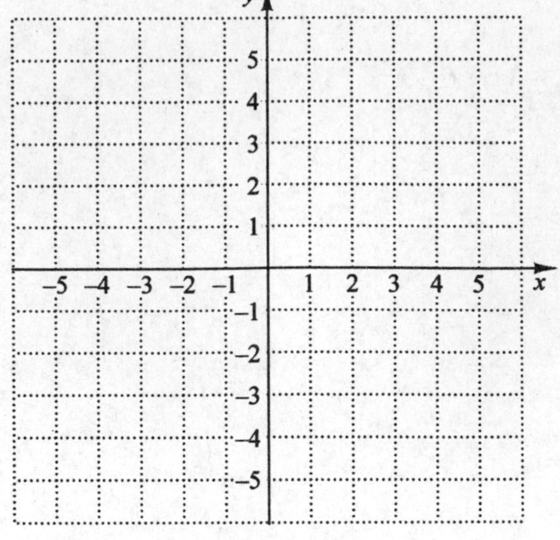

13. $y \ge -3$,
$y \le 2 - x$

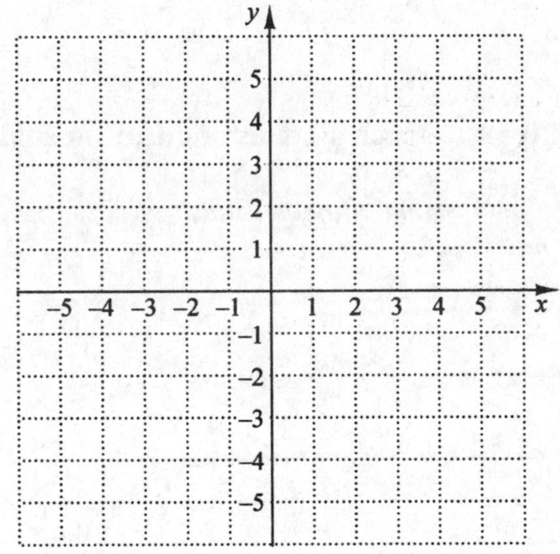

14. $2x + y \ge 1$,
$2x + y \le 4$

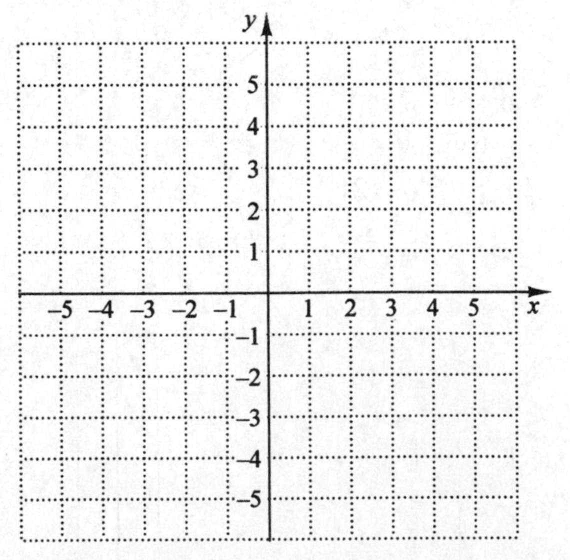

Name: Date:
Instructor: Section:

15. $y \geq x-6$,
$y \leq 3x-4$,
$x \leq 5$

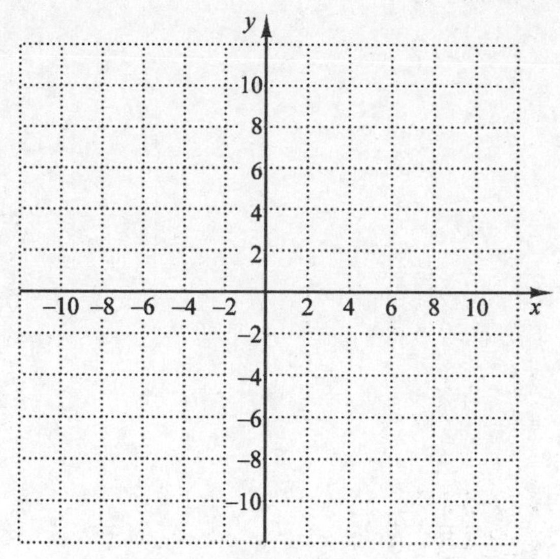

16. $2x+4y \leq 24$,
$5x+3y \leq 30$,
$x \geq 0$,
$y \geq 0$

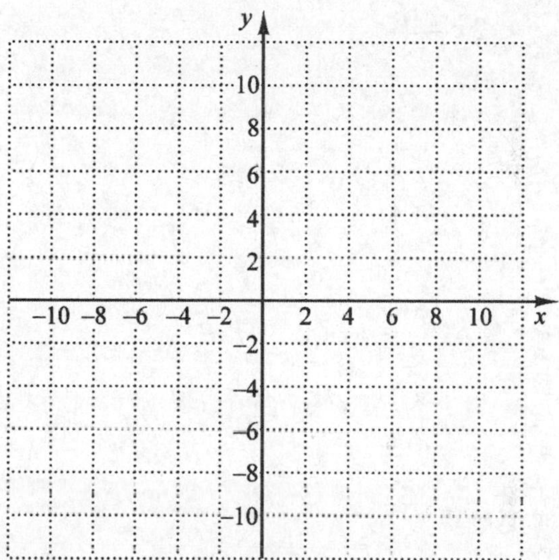

17. $-4 > x-2y$,
$x+6y < 36$,
$2y \geq -3x-12$

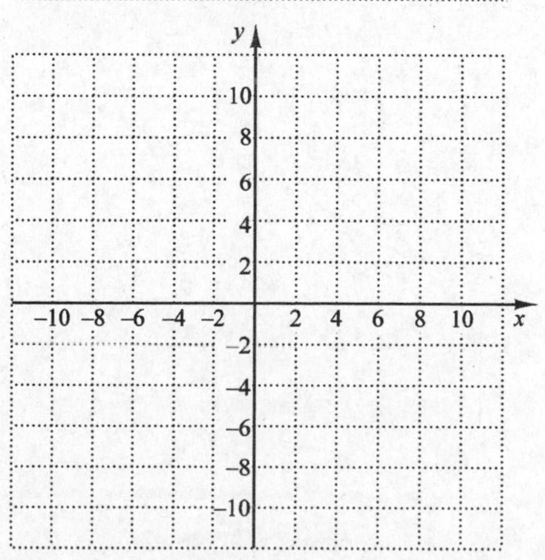

18. $y-4>2x$,
$y>-6x$,
$3y-18>x$

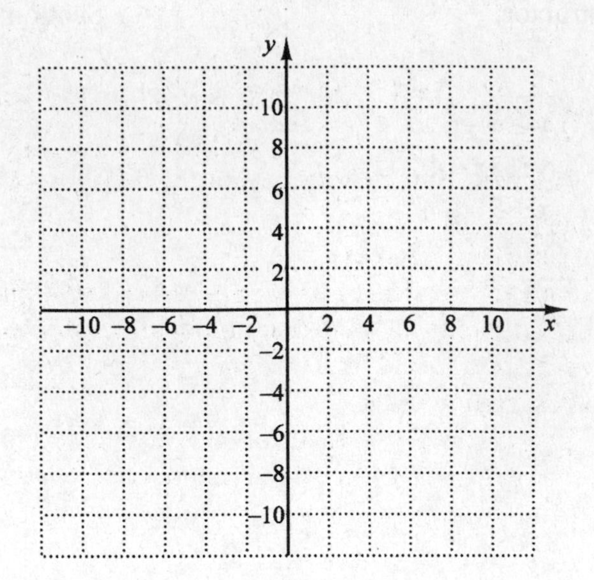

19. 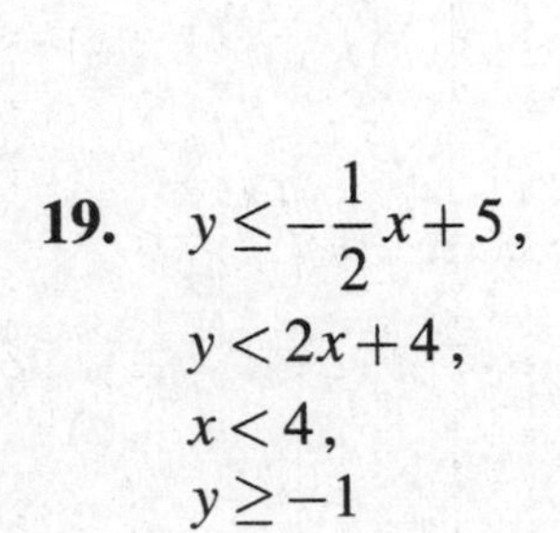
$y\leq-\frac{1}{2}x+5$,
$y<2x+4$,
$x<4$,
$y\geq-1$

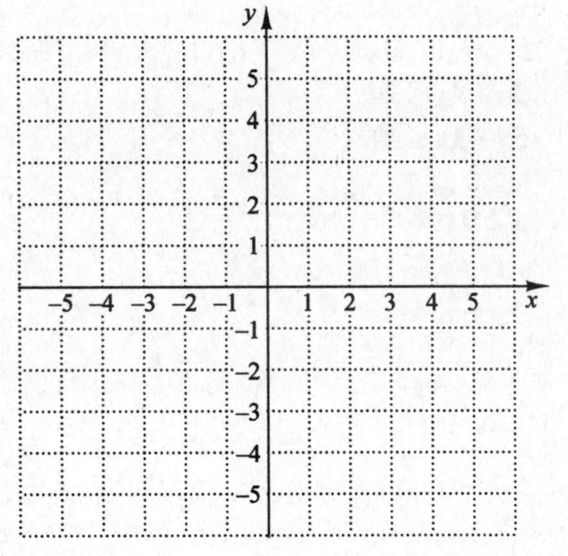

Name: Date:
Instructor: Section:

Chapter 4 POLYNOMIALS AND POLYNOMIAL FUNCTIONS

4.1 Introduction to Polynomials and Polynomial Functions

Learning Objectives

a Identify the degree of each term and the degree of a polynomial; identify terms, coefficients, monomials, binomials, and trinomials; identify the leading term, the leading coefficient, and the constant term; and arrange polynomials in ascending order or descending order.
b Evaluate a polynomial function for given inputs.
c Collect like terms in a polynomial and add polynomials.
d Find the opposite of a polynomial and subtract polynomials.

Key Terms

Use the vocabulary terms listed below to complete each statement in Exercises 1–4.

term **coefficient** **degree** **polynomial**

1. A ________________________ is a monomial or a sum of monomials.

2. A polynomial composed of two ________________________(s) is a binomial.

3. The ________________________ of a term is the number of variable factors in that term.

4. The constant factor of a term is the ________________________ of that term.

Objective a Identify the degree of each term and the degree of a polynomial; identify terms, coefficients, monomials, binomials, and trinomials; identify the leading term, the leading coefficient, and the constant term; and arrange polynomials in ascending order or descending order.

Arrange in descending powers of y.

5. $11-6b^5+3b-5b^3$ **5.** ________________

6. $r+2r^4-r^7-19r^3+3r^5$ **6.** ________________

Arrange in ascending powers of x.

7. $3x+8+2x^8-3x^6$ **7.** ________________

8. $-9x^3y+8xy^2+x^2y^3+2x^5$ **8.** ________________

For the polynomial given, answer the following questions.
a) Identify the terms.
b) What is the degree of each term?
c) What is the degree of the polynomial?
d) What is the leading term?
e) What is the leading coefficient?
f) What is the constant term?

9. $-4w^6-7w^5+5w^4+5w-9$

9a. ________________

b. ________________

c. ________________

d. ________________

e. ________________

f. ________________

10. $-5x^7+2x^5-3x^3+4x^2+6x+11$

10a. ________________

b. ________________

c. ________________

d. ________________

e. ________________

f. ________________

Name: Date:
Instructor: Section:

Objective b Evaluate a polynomial function for given inputs.

11. *Find* $g(3)$ *for the polynomial function* $g(x) = x^3 - 5x^2 + x$

11. ________________

12. *Find* $f(-2)$ *for the polynomial function* $f(x) = -4x^3 + 9x^2 - 3x - 2$

12. ________________

Evaluate each polynomial function for the given values of the variable.

13. Find $P(6)$ and $P(0)$: $P(v) = 6v^2 - 11v + 5$

13. ________________

14. Find $Q(-4)$ and $Q\left(\frac{1}{4}\right)$: $Q(b) = 4b^3 - 9b - 2$

14. ________________

Solve.

15. For a club consisting of n people, the number of ways in which a president, vice president, and treasurer can be elected can be determined using the function $p(n) = n^3 - 3n^2 + 2n$. Find the number if the club has 17 members.

15. ________________

16. The distance s, in feet, traveled by a body falling freely from rest in t seconds is approximated by the polynomial $s = 16t^2$. A brick is dropped from the top of a building and takes 1 second to hit the ground. How tall is the building?

16. ________________

17. In a psychology experiment, participants were able to memorize an average of M words in t minutes, where $M = -0.0011t^3 + 0.07t^2$. Use the graph of M to estimate the number of words memorized after 12 minutes Round to the nearest integer.

17. 

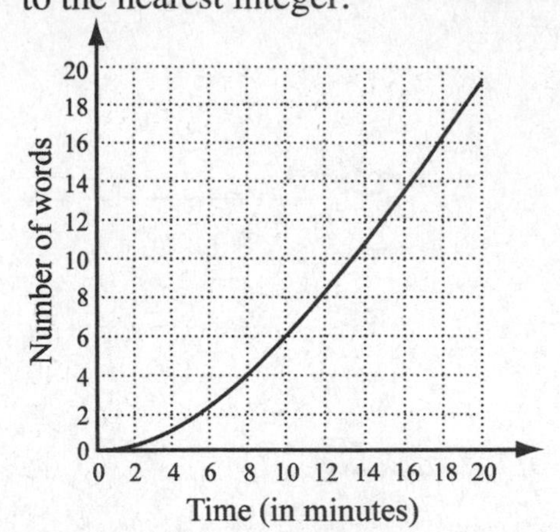

18. The polynomial function $M(t) = 0.5t^4 + 3.45t^3 - 96.65t^2 + 347.7t$, $0 \leq t \leq 6$, estimates the number of milligrams of ibuprofen in the bloodstream t hours after 400 mg of medication has been swallowed. Use the graph to estimate the number of milligrams of ibuprofen in the bloodstream 4 hours after 400 mg has been swallowed. Round to the nearest 40 mg.

18. ________________

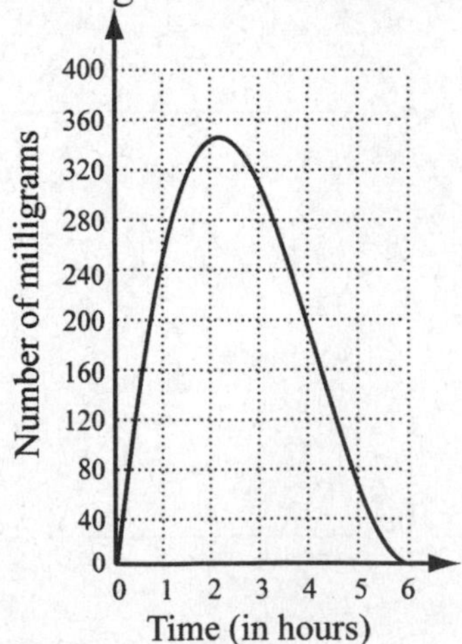

Name: Date:
Instructor: Section:

Objective c Collect like terms in a polynomial and add polynomials.

Collect like terms.

19. $5x^5 - 5x^6 - 4x^5 + 6x^6$

19. ______________

20. $-7x + 6x^2 - 2x + 4x^2 + 3$

20. ______________

21. $3x + 2x + 3x - x^5 - 8x^5$

21. ______________

22. $-x + \frac{3}{4} + 27x^7 - x - \frac{1}{2} - 2x^7$

22. ______________

Add.

23. $(7x + 5) + (-8x + 2)$

23. ______________

24. $(-4x + 8) + (x^2 + x - 9)$

24. ______________

25. $(1.2x^3 + 4.2x^2 - 4.5x) + (-3.1x^3 - 4.5x^2 + 62)$

25. ______________

26. $(7 + 3x + 4x^2 + 4x^3) + (6 - 3x + 4x^2 - 4x^3)$

26. ______________

27. $\left(\frac{1}{12}x^4 + \frac{1}{4}x^3 + \frac{3}{8}x^2 + 5\right) + \left(-\frac{7}{12}x^4 + \frac{5}{8}x^2 - 5\right)$

27. ______________

Objective d Find the opposite of a polynomial and subtract polynomials.

Write two equivalent expressions for the opposite of the polynomial.

28. $-x^2+16x-3$ **28.** ______________

Subtract.

29. $(-6x+9)-(x^2+x-8)$ **29.** ______________

30. $(7x^4+7x^3-2)-(9x^2-2x+5)$ **30.** ______________

31. $(1.4x^3+4.2x^2-3.3x)-(-4.4x^3-4.2x^2+92)$ **31.** ______________

32. $(4n^3+7n)-(-9n^3-6n+2)$ **32.** ______________

33. $\left(\frac{1}{6}x^3-\frac{3}{8}x-\frac{1}{6}\right)-\left(-\frac{1}{6}x^3+\frac{3}{8}x-\frac{1}{6}\right)$ **33.** ______________

34. $(0.08x^3-0.05x^2+0.05x)-(0.07x^3+0.08x^2-5)$ **34.** ______________

Name: Date:
Instructor: Section:

Chapter 4 POLYNOMIALS AND POLYNOMIAL FUNCTIONS

4.2 Multiplication of Polynomials

Learning Objective

a Multiply any two polynomials.
b Use the FOIL method to multiply two binomials.
c Use a rule to square a binomial.
d Use a rule to multiply a sum and a difference of the same two terms.
e For functions f described by second-degree polynomials, find and simplify notation like $f(a+h)$ and $f(a+h)-f(a)$.

Objective a Multiply any two polynomials.

Multiply.

1. $\left(2x^9\right)(6)$ **1.** ____________

2. $\left(-x^9\right)(-x)$ **2.** ____________

3. $x^3(x^5+1)$ **3.** ____________

4. $\left(-4x^5\right)\left(x^5+x\right)$ **4.** ____________

5. $(x+4)(x+2)$ **5.** ____________

6. $(x+6)(x-3)$ **6.** ____________

7. $\left(x^2+x+7\right)(x-7)$ **7.** ____________

8. $(4x+2)(2x^2+3x+1)$

8. ______________

9. $(y^2-8)(4y^2-9y+2)$

9. ______________

10. $(-8x^3-5x^2+2)(5x^2-x)$

10. ______________

11. $(x^2+x+3)(x^2-x-3)$

11. ______________

12. $(6t^2-t-6)(9t^2+6t-1)$

12. ______________

Objective b **Use the FOIL method to multiply two binomials.**
Objective c **Use a rule to square a binomial.**

Multiply.

13. $(x+8)^2$

13. ______________

14. $(2x^2+2)^2$

14. ______________

15. $\left(a-\frac{9}{2}\right)^2$

15. ______________

16. $(5-5x^8)^2$

16. ______________

Name: Date:
Instructor: Section:

17. $\left(9-5x^3\right)^2$ **17.** ______________

18. $(5x+6h)^2$ **18.** ______________

19. $\left(r^9t^7-9\right)^2$ **19.** ______________

Objective d **Use a rule to multiply a sum and a difference of the same two terms.**

Multiply.

20. $(4t-5)(4t+5)$ **20.** ______________

21. $\left(p-\frac{1}{5}\right)\left(p+\frac{1}{5}\right)$ **21.** ______________

22. $(x+7)(x-7)$ **22.** ______________

23. $\left(6x^9+3\right)\left(6x^9-3\right)$ **23.** ______________

24. $\left(x^5-x^2\right)\left(x^5+x^2\right)$ **24.** ______________

25. $\left(v^2+tj\right)\left(v^2-tj\right)$ **25.** ______________

26. $(x+y-12)(x+y+12)$

26. ______________

27. $[x+y+1][x-(y+1)]$

27. ______________

28. $\left(-bc+a^2\right)\left(bc+a^2\right)$

28. ______________

Objective e For functions *f* described by second-degree polynomials, find and simplify notation like *f* (*a* + *h*) and *f* (*a* + *h*) − *f* (*a*).

29. Given $f(x)=5x+x^2$, find and simplify each of the following.
a) $f(t-7)$

b) $f(w+h)-f(w)$

29.
a. ______________

b. ______________

30. Given $f(x)=4x^2-6x+3$, find and simplify each of the following.
a) $f(t-5)$

b) $f(a+h)-f(a)$

30.
a. ______________

b. ______________

31. Given $f(x)=8-5x-2x^2$, find and simplify each of the following.
a) $f(p+2)$

b) $f(w+h)-f(w)$

31.
a. ______________

b. ______________

Name: Date:
Instructor: Section:

Chapter 4 POLYNOMIALS AND POLYNOMIAL FUNCTIONS

4.3 Introduction to Factoring

Learning Objectives
a Factor polynomials whose terms have a common factor.
b Factor certain polynomials with four terms by grouping.

Objective a Factor polynomials whose terms have a common factor.

Factor.

1. $y^4 + 3y^3$ **1.** ______________

2. $x^2 - 8x$ **2.** ______________

3. $4t^3 - 20t^2$ **3.** ______________

4. $6x^2y - 10xy^2$ **4.** ______________

5. $6x^3y^4 - 18x^2y^5$ **5.** ______________

6. $7a^2 - 14a + 49$ **6.** ______________

7. $15x^5yz^5 - 10x^2y^4z^6 + 25x^3y^3z^4$ **7.** ______________

Factor out a common factor with a negative coefficient.

8. $-10x+40$

8. ______________

9. $-6x^2+36x+48$

9. ______________

10. $-3a+3b$

10. ______________

11. $-2y^4+12s^3$

11. ______________

12. $-r^5+5r^4-20r$

12. ______________

13. $-t^4-t^3+2t+13$

13. ______________

14. A hobby rocket is launched upward with an initial velocity of 64 ft/sec. Its height in feet, $h(t)$, after t seconds is given by $h(t)=-16t^2+64t$.

a) Find an equivalent expression for $h(t)$ by factoring out a common factor with a negative coefficient.

b) Check your factoring by evaluating both expressions for $h(t)$ at $t=1$.

14.
a)______________

b)______________

Name: Date:
Instructor: Section:

15. When x storage sheds are sold, More Space collects a profit of $P(x)$, where $P(x) = 10x^2 - 9x$ (in hundreds of dollars). Find an equivalent expression by factoring out a common factor.

15. ______________

Objective b Factor certain polynomials with four terms by grouping.

Factor.

16. $x(y+2)+z(y+2)$

16. ______________

17. $(p-2)(p+4)+(p-2)(p+6)$

17. ______________

18. $6t^3(t-1)+5(1-t)$

18. ______________

19. $uv+vw+tu+tw$

19. ______________

20. $p^2(y-r)+5(r-y)$

20. ______________

21. $x^3 - 5x^2 - 2x + 10$ **21.** ______________

22. $8m^5 + 20m^4 - 24m^3 - 28m^2$ **22.** ______________

23. $a^5 - a^4 - 5a^2 + 5a^3$ **23.** ______________

24. $5pq - p^2q + 4p - 20$ **24.** ______________

25. $5s^4 + 15s^2 + 7s^2 + 21$ **25.** ______________

Name: Date:
Instructor: Section:

Chapter 4 POLYNOMIALS AND POLYNOMIAL FUNCTIONS

4.4 Factoring Trinomials: $x^2 + bx + c$

Learning Objectives
a Factor trinomials of the type $x^2 + bx + c$.

Objective a Factor trinomials of the type $x^2 + bx + c$.

Factor.

1. $x^2 + 11x + 18$ **1.** ____________

2. $t^2 - 13t + 40$ **2.** ____________

3. $x^3 - 4x^2 - 12x$ **3.** ____________

4. $44 + a^2 - 15a$ **4.** ____________

5. $t^2 + 8t + 11$ **5.** ____________

6. $15x + 36 + x^2$ **6.** ____________

7. $42 - n - n^2$ **7.** ____________

8. $3x - 18 + x^2$ **8.** ________________

9. $a^2 - 9ab - 22b^2$ **9.** ________________

10. $y^5 - 74y^4 + 73y^3$ **10.** ________________

11. $x^4 + 82x^2 + 81$ **11.** ________________

12. $x^6 - 10x^3 + 21$ **12.** ________________

13. $x^8 - 7x^4 - 30$ **13.** ________________

14. $y^2 + 0.1y - 0.12$ **14.** ________________

15. $9 - 8t^{12} - t^{24}$ **15.** ________________

16. $36 - a^2 - 9a$ **16.** ________________

Name: Date:
Instructor: Section:

Chapter 4 POLYNOMIALS AND POLYNOMIAL FUNCTIONS

4.5 Factoring Trinomials: $ax^2 + bx + c, a \neq 1$

Learning Objective
a Factor trinomials of the type $ax^2 + bx + c$, $a \neq 1$, by the FOIL method.
b Factor trinomials of the type $ax^2 + bx + c$, $a \neq 1$, by the *ac*- method.

Objective a Factor trinomials of the type $ax^2 + bx + c$, $a \neq 1$, by the FOIL method.
Objective b Factor trinomials of the type $ax^2 + bx + c$, $a \neq 1$, by the *ac*- method.

Factor.

1. $5n^2 + 25n - 70$ **1.** ______________

2. $12z^3 - 24z^2 - 180z$ **2.** ______________

3. $10x^2 + x - 3$ **3.** ______________

4. $6a^3 - 11a^2 - 10a$ **4.** ______________

5. $40t^2 - 38t - 15$ **5.** ______________

6. $5x - 15 + 20x^2$ **6.** ______________

7. $35m^5 + 9m^4 - 2m^3$ 7. ______________

8. $4x^2 + 16x + 15$ 8. ______________

9. $-18z^2 + 24z + 10$ 9. ______________

10. $12 - 20x - 25x^2$ 10. ______________

11. $8a^3b - 2a^2b - 15ab$ 11. ______________

12. $22x^2 - 4 + 87x$ 12. ______________

13. $30t^4 - 63t^3 - 30t^2$ 13. ______________

14. $6p^2 - 19pq + 15q^2$ 14. ______________

15. $25c^2 - 20cd + 4d^2$ 15. ______________

16. $8a^2b^2 + 13ab - 6$ 16. ______________

Name: Date:
Instructor: Section:

Chapter 4 POLYNOMIALS AND POLYNOMIAL FUNCTIONS

4.6 Special Factoring

Learning Objective
a Factor trinomial squares.
b Factor differences of squares.
c Factor certain polynomials with four terms by grouping and possibly using the factoring of a trinomial square or the difference of squares.
d Factor sums and differences of cubes.

Key Terms
Use the vocabulary terms listed below to complete each statement in Exercises 1–3.

difference of squares **factored completely** **perfect-square trinomial**

1. An expression that cannot be factored further is ___________________.

2. The expression $x^2 - 36$ is an example of a ___________________.

3. The expression $x^2 - 12x + 36$ is an example of a ___________________.

In Exercises 4–6, match the term with the appropriate example in the column on the right.

4. _____ difference of cubes a) $t^2 - s^2$

5. _____ difference of squares b) $5^3 - 8^3$

6. _____ sum of cubes c) $z^3 + 10^3$

Objective a Factor trinomial squares.

Factor.

7. $v^2 - 6v + 9$ 7. _______________

8. $x^2 - 2x + 1$ 8. _______________

9. $y^2 + 16 + 8y$ 9. _______________

10. $2a^2 + 12a + 18$ — 10. ______________

11. $25 - 10n + n^2$ — 11. ______________

12. $y^2 + 4xy + 4x^2$ — 12. ______________

13. $100x + x^3 + 20x^2$ — 13. ______________

14. $0.01c^2 - 0.10c + 0.25$ — 14. ______________

Objective b Factor differences of squares.

Factor.

15. $t^2 - 1$ — 15. ______________

16. $9 - m^2n^2$ — 16. ______________

17. $5ab^2 - 20a$ — 17. ______________

18. $15x^4 - 15y^4$ — 18. ______________

19. $64x^2y^8 - 25x^6$ — 19. ______________

20. $\frac{1}{4} - y^2$ — 20. ______________

Name: Date:
Instructor: Section:

Objective c Factor certain polynomials with four terms by grouping and possibly using the factoring of a trinomial square or the difference of squares.

Factor.

21. $(y-z)^2-64$ **21.** ____________

22. x^3+x^2-9x-9 **22.** ____________

23. $p^2+10pq+25q^2-100t^2$ **23.** ____________

24. $16w^4-25w^2d^4$ **24.** ____________

25. $\frac{1}{25}-w^2$ **25.** ____________

26. $(v+d)^2-25$ **26.** ____________

27. $w^2-2w+1-25p^2$ **27.** ____________

28. $c^3 - 9c^2 - 4c + 36$

28. ________________

29. $5w^3 - 5w^2y - 5wy^2 + 5y^3$

29. ________________

30. $s^2 + 2sq + q^2 - 81$

30. ________________

31. $4 - (x + y)^2$

31. ________________

32. $a^2 - 2a + 1 - 81p^2$

32. ________________

33. $64 - (w^2 + 2wt + t^2)$

33. ________________

Objective d Factor sums and differences of cubes.

Factor.

34. $x^3 + 125$

34. ________________

35. $64t^3 - 1$

35. ________________

Name: Date:
Instructor: Section:

36. $8+27y^3$ **36.** ______________

37. z^3+w^3 **37.** ______________

38. $6-6p^3$ **38.** ______________

39. $d^3-0.008$ **39.** ______________

40. $3a^3+24b^3$ **40.** ______________

41. $x^3-\frac{1}{27}$ **41.** ______________

42. pq^3-p^4 **42.** ______________

43. $1250t^6+10s^6$ **43.** ______________

44. $3m^5 + 3000m^2$

44. ______________

45. $x^6 - 1$

45. ______________

46. $y^6 - 64z^6$

46. ______________

47. $x^{15} + x^6y^{12}$

47. ______________

48. $8b^3 + 27$

48. ______________

49. $r^6 - 64$

49. ______________

50. $a^{12} + b^{15}c^6$

50. ______________

Name: Date:
Instructor: Section:

Chapter 4 POLYNOMIALS AND POLYNOMIAL FUNCTIONS

4.7 Factoring: A General Strategy

Learning Objectives
a Factor polynomials completely using any of the methods considered in this chapter.

Key Terms
Use the vocabulary terms listed below to complete each statement in Exercises 1–4.

common factor	**difference of squares**
factor completely	**grouping**

1. When factoring a polynomial, always look first for a(n) ____________________.

2. If there are two terms in a polynomial, try first to factor as a(n) ____________________.

3. If there are four terms in a polynomial, try to factor by ____________________.

4. Always ____________________.

Objective a Factor polynomials completely using any of the methods considered in this chapter.

Factor completely.

5. $2x^2 - x - 6$ **5.** ________________

6. $3y^2 - 12$ **6.** ________________

7. $a^3 - 12a^2 + 36a$ **7.** ________________

8. $x^2 + 81 + 18x$

8. ________________

9. $x^3 - 2x^2 - 9x + 18$

9. ________________

10. $8x^3 - 50x$

10. ________________

11. $6t^3 + 26t^2 - 20t$

11. ________________

12. $12x^2 - 80x - 28$

12. ________________

13. $-3y^5 + 24y^4 - 48y^3$

13. ________________

14. $c^4 - 25$

14. ________________

15. $3t^6 - 48t^2$

15. ________________

16. $m^3 - n^3$

16. ________________

Name: Date:
Instructor: Section:

17. c^2x+c^2y **17.** ______________

18. $14p^2q-7pq$ **18.** ______________

19. $2x(5x-y)-(5x-y)$ **19.** ______________

20. $r^2+5r+rs+5s$ **20.** ______________

21. $t^2-4t+4-100v^2$ **21.** ______________

22. $2x^2+xy-15y^2$ **22.** ______________

23. $9p^2-12pq+4q^2$ **23.** ______________

24. $a^2+12ab+36b^2$ **24.** ______________

25. y^6-64 **25.** ______________

26. $2x^3 - 6x^2y + 5xy^2$

26. ________________

27. $20m^2 - 7mn - 6n^2$

27. ________________

28. $-x^2 - 5x + 24$

28. ________________

29. $a^4 - b^6 - 10b^3 - 25$

29. ________________

30. $9t^2 - 8t + \frac{16}{9}$

30. ________________

31. $\frac{1}{9}a^2 + \frac{4}{15}a + \frac{4}{25}$

31. ________________

32. $x^4 + 5x^3 + 5x^2 + 25x$

32. ________________

Name: Date:
Instructor: Section:

Chapter 4 POLYNOMIALS AND POLYNOMIAL FUNCTIONS

4.8 Applications of Polynomial Equations and Functions

Learning Objectives

a Solve quadratic and other polynomial equations by first factoring and then using the principle of zero products.

b Solve applied problems involving quadratic and other polynomial equations that can be solved by factoring.

Key Terms

Use the vocabulary terms listed below to complete each statement in Exercises 1–4.

hypotenuse **leg** **Pythagorean theorem** **right triangle**

1. A(n) ____________________ contains a 90° angle.

2. The ____________________ of a right triangle is the side opposite the right angle.

3. A ____________________ of a right triangle forms one of the sides of the right angle.

4. The ____________________ states that the sum of the squares of the lengths of the legs of a right triangle is equal to the square of the length of the hypotenuse.

Objective a Solve quadratic and other polynomial equations by first factoring and then using the principle of zero products

Solve.

5. $3x(x-9)=0$ **5.** ________________

6. $x^2+12x=0$ **6.** ________________

7. $-12x^2=24x$ **7.** ________________

8. $6x^4 = 8x^5$

8. ________________

9. $x^2 - 6x = 16$

9. ________________

10. $n^3 + n^2 - 6n = 0$

10. ________________

11. $a^2 = 12a$

11. ________________

12. $50 - x^2 = 23x$

12. ________________

13. $3x^2 + 2x - 5 = 0$

13. ________________

Name: Date:
Instructor: Section:

14. $3x^3 - 11x^2 - 4x = 0$

14. ________________

15. $16x^2 + 34x = 15$

15. ________________

16. $2x(4x+5) = 7$

16. ________________

Find the domain of the function f given by each of the following.

17. $f(x) = \dfrac{5}{5x - 35x^2}$

17. ________________

18. $f(x) = \dfrac{x+4}{x^3 + x^2 - 6x}$

18. ________________

Objective b Solve applied problems involving quadratic and other polynomial equations that can be solved by factoring.

Solve.

19. A scrapbook is 3 in. longer than it is wide. Find the length and the width if the area is 108 in^2.

19. ______________

20. A picture frame measures 16 cm by 20 cm, and 192 cm^2 of picture shows. Find the width of the frame.

20. ______________

21. A rectangular playground is 100 ft by 70 ft. Part of the playground is torn up to plant trees in a strip of uniform width around the playground. The area of the new playground is 4000 ft^2. How wide is the strip of trees?

21. ______________

22. Three consecutive odd integers are such that the square of the first plus the square of the third is 170. Find the three integers.

22. ______________

Name: Date:
Instructor: Section:

23. A wire is stretched from the ground to the top of a pole. The wire is 34 ft long. The height of the pole is 14 ft greater than the distance d from the pole's base to the bottom of the wire. Find the distance d and the height of the pole.

23. ______________

24. A triangular sign is 2 ft taller than it is wide. The area is 24 ft^2. Find the height and the base of the sign.

24. ______________

25. A rectangular yard is 20 ft longer than it is wide. Determine the dimensions of the yard if it measures 100 ft diagonally.

25. ______________

26. The foot of an extension ladder is 15 ft from a wall. The ladder is 5 ft longer than the height that it reaches on the wall. How far up the wall does the ladder reach?

26. ______________

27. A flower bed is to be 8 m longer than it is wide. The flower bed will have an area of 84 m^2. What will its dimensions be?

27. ________________

28. The height $h(t)$, in feet, of a flare launched upward with an initial velocity of 48 ft/sec from a height of 160 ft after t seconds can be approximated by $h(t) = -16t^2 + 48t + 160$. After how long will the flare reach the ground?

28. ________________

Name: Date:
Instructor: Section:

Chapter 5 RATIONAL EXPRESSIONS, EQUATIONS, AND FUNCTIONS

5.1 Rational Expressions and Functions: Multiplying, Dividing, and Simplifying

Learning Objectives

a Find all numbers for which a rational expression is not defined or that are not in the domain of a rational function, and state the domain of the function.
b Multiply a rational expression by 1, using an expression like *A/A*.
c Simplify rational expressions.
d Multiply rational expressions and simplify.
e Divide rational expressions and simplify.

Key Terms

Use the vocabulary terms listed below to complete each statement in Exercises 1–3.

rational expression **simplify** **invert**

1. To ________________________ a rational expression, factor and remove factors equal to 1.

2. A polynomial divided by a nonzero polynomial is called a ____________________.

3. When we divide rational expressions, we often say that we __________________ and multiply.

Objective a Find all numbers for which a rational expression is not defined or that are not in the domain of a rational function, and state the domain of the function.

Find all numbers for which the rational expression is not defined.

4. $\frac{67x}{x^2+x-20}$ **4.** ________________

5. $\frac{x^2-7}{x^2+2x-35}$ **5.** ________________

Find the domain. Write both set-builder notation and interval notation for the answer.

6. $f(x)=\frac{4x^2-5x+3}{x+1}$ **6.** ________________

7. $f(x)=\frac{-6x^2+2x+1}{x-1}$ **7.** ________________

8. $f(x)=\dfrac{5x+1}{x^2-x-6}$ 8. ______________

9. $f(x)=\dfrac{x^2-x}{x^2-16}$ 9. ______________

Objective b Multiply a rational expression by 1, using an expression like *A/A*.

Multiply to obtain an equivalent expression. Do not simplify.

10. $\dfrac{2x}{2x}\cdot\dfrac{x+9}{x+4}$ 10. ______________

11. $\dfrac{y-6}{y+1}\cdot\dfrac{y-7}{y-7}$ 11. ______________

Objective c Simplify rational expressions.

Simplify.

12. $\dfrac{24r^{10}}{32r^{12}}$ 12. ______________

13. $\dfrac{30a^8b^5}{12ab^4}$ 13. ______________

14. $\dfrac{5a-10}{5}$ 14. ______________

15. $\dfrac{8m-40}{8m}$ 15. ______________

16. $\dfrac{4y-28}{4y+28}$ 16. ______________

17. $\dfrac{a^2+a-2}{a^2+2a-3}$ 17. ______________

18. $\dfrac{2x^2+14x+20}{6x^2+12x-90}$ 18. ______________

Name: Date:
Instructor: Section:

Objective d Multiply rational expressions and simplify.

Multiply and simplify.

19. $\frac{4t}{6t+9}\cdot\frac{8t+12}{t^3}$ **19.** ____________

20. $\frac{a+b}{a-b}\cdot\frac{8a-8b}{a^2-b^2}$ **20.** ____________

21. $\frac{x^2+1}{x^2-5x+6}\cdot\frac{x+1}{x-1}$ **21.** ____________

22. $\frac{x^2+6x+5}{x^2-4x+3}\cdot\frac{x-3}{x+5}$ **22.** ____________

23. $\frac{3y^2-27}{2y^2-128}\cdot\frac{8y+64}{6y-6}$ **23.** ____________

24. $\frac{a^4-a}{4a-8}\cdot\frac{4a^2-16}{2a^4+2a^3+2a^2}$ **24.** ____________

Objective e Divide rational expressions and simplify.

Divide and simplify.

25. $\frac{x^2}{y^3}\div\frac{x^5}{y}$ **25.** ____________

26. $\frac{t^2-4}{t^2}\div\frac{t+2}{t-2}$ **26.** ____________

27. $\dfrac{5a+5}{a-3} \div \dfrac{a+1}{a-7}$ **27.** ____________

28. $\dfrac{x^2-1}{x^2+1} \div \dfrac{x+1}{x-1}$ **28.** ____________

29. $\dfrac{x^2+4x+4}{3x-9} \div \dfrac{x^2-3x-10}{x^2-9}$ **29.** ____________

30. $\dfrac{4x^2-8x-5}{2x^2-11x-21} \div \dfrac{6x^2-13x-5}{4x^2+8x+3}$ **30.** ____________

Perform the indicated operations and simplify.

31. $\left(\dfrac{8k^2-6k-9}{k^2+6k} \div \dfrac{2k-3}{k^3+3k^2-18k}\right) \cdot \dfrac{k^2-5k+6}{4k^2-5k-6}$ **31.** ____________

32. $\dfrac{a^2+10a+24}{6a^2-35a-6} \cdot \dfrac{3a+b}{a+3b} \div \dfrac{a+6}{a-6}$ **32.** ____________

Name: Date:
Instructor: Section:

Chapter 5 RATIONAL EXPRESSIONS, EQUATIONS, AND FUNCTIONS

5.2 LCMs, LCDs, Addition, and Subtraction

Learning Objective
a Find the LCM of several algebraic expressions by factoring.
b Add and subtract rational expressions.
c Simplify combined additions and subtractions of rational expressions.

Key Terms
Use the vocabulary terms listed below to complete each statement in Exercises 1–2.

least common denominator **least common multiple**

1. The expression $12x^2y^3$ is the ________________ of $6xy^2$, $4x^2y$, and $2y^3$.

2. The expression $12x^2y^3$ is the ________________ of $\frac{5}{6xy^2}$, $\frac{1}{4x^2y}$, and $\frac{3x}{2y^3}$.

Objective a Find the LCM of several algebraic expressions by factoring.

Find the LCM by factoring.

3. 24, 30 **3.** ____________

4. 8, 15 **4.** ____________

5. 6, 15, 20 **5.** ____________

6. 10, 50, 120 **6.** ____________

Add. Find the LCD first.

7. $\frac{5}{12}+\frac{2}{15}$ **7.** ____________

8. $\frac{5}{18}+\frac{11}{24}$ **8.** ____________

9. $\frac{3}{15}+\frac{7}{20}+\frac{2}{25}$ **9.** ____________

Find the LCM.

10. $10x^3,\ 30x^5$

10. ______________

11. $3y^3,\ 12x^2y,\ 15xy^4$

11. ______________

12. $a^2-9,\ a^2-2a-3$

12. ______________

13. $m^3+6m^2+9m,\ m^2-3m$

13. ______________

14. $x^2-x-30,\ x^2-7x+6$

14. ______________

15. $4y^5-24y^4+36y^3,\ 6y^3+12y^2-90y$

15. ______________

Objective b Add and subtract rational expressions.

Add or subtract. Then simplify. If a denominator has three or more factors (other than monomials), leave it in factored form.

16. $\dfrac{4x}{x+5}+\dfrac{2x-3}{x+5}$

16. ______________

17. $\dfrac{y^2}{y-10}+\dfrac{y-110}{y-10}$

17. ______________

18. $\dfrac{3}{x}-\dfrac{9}{-x}$

18. ______________

Name: Date:
Instructor: Section:

19. $\frac{5-t}{t-4}-\frac{3t-2}{4-t}$

19. ______________

20. $\frac{x+1}{x-4}+\frac{x-4}{x+1}$

20. ______________

21. $\frac{2}{x^2-6x-7}+\frac{5}{x^2-2x-3}$

21. ______________

22. $\frac{4}{x^2+x-20}-\frac{3}{x^2-25}$

22. ______________

23. $\frac{x}{x^2+4x+3}-\frac{1}{x^2-1}$

23. ______________

Objective c Simplify combined additions and subtractions of rational expressions.

Perform the indicated operations and simplify.

24. $\frac{x+5}{x-3}-\frac{2-x}{x+3}-\frac{4x-18}{9-x^2}$

24. ______________

25. $\frac{3x}{1-4x}+\frac{2x}{4x+1}-\frac{1}{16x^2-1}$

25. ______________

26. $\frac{1}{a-b}+\frac{2}{a-b}+\frac{2a}{a^2-b^2}$

26. ______________

27. $\frac{12y}{y^2-36}-\frac{6}{y}-\frac{6}{y+6}$

27. ______________

28. $\frac{7w}{w^2-1}+\frac{2w}{1-w}-\frac{7}{w-1}$

28. ______________

Name: Date:
Instructor: Section:

Chapter 5 RATIONAL EXPRESSIONS, EQUATIONS, AND FUNCTIONS

5.3 Division of Polynomials

Learning Objectives

a Divide a polynomial by a monomial.
b Divide a polynomial by a divisor that is not a monomial, and if there is a remainder, express the result in two ways.
c Use synthetic division to divide a polynomial by a binomial of the type $x - a$.

Key Terms

Use the vocabulary terms listed below in Exercises 1–4 to label each part of the division shown.

dividend **divisor** **quotient** **remainder**

1. ____________________
2. ____________________
3. ____________________
4. ____________________

$$\begin{array}{r} x + 2 \;\leftarrow (1) \\ (2) \rightarrow x+3 \,\overline{)\, x^2 + 5x + 10} \;\leftarrow (3) \\ \underline{x^2 + 3x } \\ 2x + 10 \\ \underline{2x + 6} \\ 4 \;\leftarrow (4) \end{array}$$

Objective a Divide a polynomial by a monomial.

Divide and check.

5. $\dfrac{7t - t^5 + 3t^6}{t}$

5. ____________________

6. $(6x^5 - 30x^4 + 12x) \div (3x)$

6. ____________________

7. $(15y^5 - 5y^3 - 30y) \div (-5y^2)$

7. ____________________

8. $\dfrac{24x^7 - 40x^5 - 18x^3}{4x^3}$

8. ______________

9. $\dfrac{36s^5t^2 - 60s^7t^3 + 42s^4t^4}{-6s^4t^2}$

9. ______________

Objective b Divide a polynomial by a divisor that is not a monomial, and if there is a remainder, express the result in two ways.

Divide.

10. $(x^2 - 2x - 80) \div (x - 10)$

10. ______________

11. $(p^2 - 4p + 12) \div (p - 3)$

11. ______________

12. $(15x^2 + 2x - 8) \div (5x + 4)$

12. ______________

Name: Date:
Instructor: Section:

13. $(2x^4 + 3x^3 - 5x^2 - x - 6) \div (x^2 - 3)$

13. ______________

14. $(2x^4 + 5x^3 - 5x^2 - 20x - 12) \div (x^2 - 4)$

14. ______________

Objective c Use synthetic division to divide a polynomial by a binomial of the type $x - a$.

Solve.

15. $(a^3 - 2a + 3) \div (a + 1)$

15. ______________

16. $(x^3 - 4x^2 + 5) \div (x - 2)$

16. ______________

17. $(x^3 + 64) \div (x + 4)$

17. ______________

18. $(x^4 - 625) \div (x - 5)$

18. ______________

19. $(x^3 - 4x^2 + 11) \div (x + 2)$

19. ______________

20. $(y^7 - 1) \div (y - 1)$

20. ______________

Name: Date:
Instructor: Section:

Chapter 5 RATIONAL EXPRESSIONS, EQUATIONS, AND FUNCTIONS

5.4 Complex Rational Expressions

Learning Objectives
a Simplify complex rational expressions.

Objective a Simplify complex rational expressions.

Solve.

1. $\dfrac{\dfrac{1}{4}+\dfrac{2}{5}}{\dfrac{3}{10}-\dfrac{4}{5}}$

1. ________________

2. $\dfrac{9-\dfrac{1}{t^2}}{3-\dfrac{1}{t}}$

2. ________________

3. $\dfrac{6+\dfrac{6}{n}}{2+\dfrac{2}{n}}$

3. ________________

4. $\dfrac{\dfrac{x}{10}-\dfrac{10}{x}}{\dfrac{1}{x}+\dfrac{1}{10}}$

4. ________________

5. $\dfrac{\dfrac{1}{p}+1}{\dfrac{1}{p^2}-1}$

5. ________________

6. $\dfrac{\dfrac{1}{2}-\dfrac{1}{x}}{\dfrac{2-x}{2}}$

6. ________________

7. $\dfrac{\dfrac{3}{y^2}-\dfrac{3}{z^2}}{\dfrac{1}{y}+\dfrac{1}{z}}$

7. ________________

8. $\dfrac{\dfrac{7}{a^3}-\dfrac{2}{a^2}}{\dfrac{5}{a^2}+\dfrac{4}{a}}$

8. ________________

Name: Date:
Instructor: Section:

9. $\dfrac{\dfrac{3}{5n^5}-\dfrac{1}{15n}}{\dfrac{6}{7n^3}+\dfrac{2}{21n}}$

9. ________________

10. $\dfrac{\dfrac{x}{y}+\dfrac{w}{z}}{\dfrac{y}{x}+\dfrac{z}{w}}$

10. ________________

11. $\dfrac{\dfrac{a}{4b^2}+\dfrac{5}{12b}}{\dfrac{5}{12b}+\dfrac{a}{4b^2}}$

11. ________________

12. $\dfrac{\dfrac{5}{n+1}+\dfrac{2}{n}}{\dfrac{3}{n+1}+\dfrac{5}{n}}$

12. ________________

13. $$\frac{\frac{x-2}{x+3}}{\frac{x+1}{x-5}}$$

13. ________________

14. $$\frac{\frac{y^2-z^2}{y}}{\frac{y+z}{yz}}$$

14. ________________

15. $$\frac{\frac{x^2-x-2}{x^2+8x+15}}{\frac{x^2-4}{x^2+6x+9}}$$

15. ________________

16. $$\frac{\frac{5}{x^2-4}+\frac{3}{x-2}}{\frac{1}{x^2-4}+\frac{3}{x+2}}$$

16. ________________

Name: Date:
Instructor: Section:

Chapter 5 RATIONAL EXPRESSIONS, EQUATIONS, AND FUNCTIONS

5.5 Solving Rational Equations

Learning Objective
a Solve rational equations.

Key Terms
Use the vocabulary terms listed below to complete each statement in Exercises 1–4.

clearing fractions **LCD** **rational equation** **solutions**

1. A(n) ________________ contains one or more rational expressions.

2. To solve a rational equation is to find all of its ________________ .

3. As a first step in solving rational equations, multiply the equation by the ________________.

4. When we multiply an equation by the least common denominator, we are ________________.

Objective a Solve rational equations.

Solve. Don't' forget to check!

5. $\frac{x}{5}+\frac{x}{4}=9$

5. ________________

6. $\frac{2}{5}-\frac{1}{t}=\frac{1}{2}$

6. ________________

7. $\frac{4}{x+2}+\frac{5}{8}=\frac{3}{2x+4}$

7. ______________

8. $\frac{3}{9}-\frac{1}{4t}=\frac{4}{12}$

8. ______________

9. $x+\frac{10}{x}=7$

9. ______________

10. $\frac{n+3}{n+1}=\frac{2}{n+1}$

10. ______________

11. $\frac{t}{t+3}=\frac{9}{t^2+3t}$

11. ______________

Name: Date:
Instructor: Section:

12. $\dfrac{2}{3x+4}=\dfrac{5}{2x}$

12. ________________

13. $\dfrac{2}{x-3}=\dfrac{x}{x+7}$

13. ________________

14. $\dfrac{20}{t-3}-\dfrac{15}{t}=\dfrac{30}{t}$

14. ________________

15. $\dfrac{5}{x-1}-\dfrac{6}{x+2}=\dfrac{3x}{x^2+x-2}$

15. ________________

16. $\dfrac{n}{n+1}-\dfrac{2}{n}=\dfrac{1}{n^2+n}$

16. ________________

17. $\dfrac{4}{x+5} - \dfrac{3}{x-5} = \dfrac{4x}{25-x^2}$

17. ________________

18. $\dfrac{4}{t^2-8t+16} + \dfrac{t+1}{3t-12} = \dfrac{t}{2t-8}$

18. ________________

For the given rational function f, find all values of x for which f(x) has the indicated value.

19. $f(x) = 2x + \dfrac{3}{x}; f(x) = 7$

19. ________________

20. $f(x) = \dfrac{x+5}{x-1}; f(x) = \dfrac{2}{3}$

20. ________________

Name: Date:
Instructor: Section:

Chapter 5 RATIONAL EXPRESSIONS, EQUATIONS, AND FUNCTIONS

5.6 Applications and Proportions

Learning Objective

a Solve work problems and certain basic problems using rational equations.
b Solve applied problems involving proportions.
c Solve motion problems using rational equations.

Key Terms

Use the variables and numbers listed below to complete each equation in Exercises 1 and 2. Variables may be used more than once.

a ***d*** ***t*** **1**

1. If a = the time needed for A to complete the work alone, b = the time needed for B to complete the work alone, and t = the time needed for A and B to complete the work together, then

$$\frac{t}{\square}+\frac{\square}{b}=\square.$$

2. Problems involving motion usually make use of the formula

$$\square=r\cdot\square.$$

Objective a Solve work problems and certain basic problems using rational equations.

Solve.

3. Mariah can clean the horse stalls at Dazzling Rides Farm in 6 hr. Lindsay needs 10 hr to complete the same job. Working together, how long will it take them to do the job?

3. ________________

4. Circle City's swimming pool can be filled in 15 hr if water enters through a pipe alone or in 24 hr if water enters through a hose alone. If water is entering through both the pipe and the hose, how long will it take to fill the pool?

4. ______________

5. Chris can clean an office in 3 hr. When he works with Hannah, they can clean the office in 80 min. How long would it take Hannah, working by herself, to clean the office?

5. ______________

6. Ethan can weed the flowerbeds by his office in 50 min. Anthony can do the same job in 45 min. How long would it take Ethan and Anthony to weed the beds if they worked together?

6. ______________

7. Abigail can file a week's worth of invoices in 75 min. Ava can do the same job in 90 min. How long would it take if they worked together?

7. ______________

Name: Date:
Instructor: Section:

Objective b Solve applied problems involving proportions.

Solve.

8. Approximately 100 cocoa beans are required to make $\frac{1}{4}$ lb of chocolate. How many beans are required to make $2\frac{1}{2}$ lb of chocolate?

8. ______________

9. Teri wrote 72 pages for her novel over a period of 12 days. At this rate, how many pages would she write in 16 days?

9. ______________

10. Linda walked 610 steps in 5 min on an elliptical trainer. At this rate, how many steps would she walk in 12 min?

10. ______________

11. The ratio of buttermilk to whole wheat flour in a flat bread recipe is $\frac{2}{3}$. If 3 cups of buttermilk are used, how many cups of whole wheat flour are used?

11. ______________

12. To determine the number of trout in his pond, Oak catches 25 trout, tags them, and lets them loose. Later, he catches 18 trout; 10 of them have tags. Assuming that the ratio of trout is the same throughout the pond, find the number of trout in the pond.

12. ______________

Objective c Solve motion problems using rational equations.

Solve.

13. The speed of the current in Pebble Creek is 4 mph. Beth can canoe 6 mi upstream in the same time it takes her to canoe 14 mi downstream. What is the speed of Beth's canoe in still water?

13. ______________

14. The speed of Tom's scooter is 16 mph less than the speed of Mary Lynne's motorcycle. The motorcycle can travel 290 mi in the same time that the scooter can travel 210 mi. Find the speed of each vehicle.

14. ______________

15. Trey's boat travels 24 km/h in still water. He motors 150 km downstream in the same time it takes to go 90 km upstream. What is the speed of the river?

15. ______________

16. A pontoon boat moves 6 km/h in still water. It travels 30 km upriver and 30 km downriver in a total time of 18 hr. What is the speed of the current?

16. ______________

Name: Date:
Instructor: Section:

Chapter 5 RATIONAL EXPRESSIONS, EQUATIONS, AND FUNCTIONS

5.7 Formulas and Applications

Learning Objectives
a Solve a formula for a letter.

Objective a Solve a formula for a letter.

Solve.

1. $\frac{1}{Y}=\frac{1}{t}+\frac{1}{m}$, for Y

1. ________________

2. $\frac{1}{100}=\frac{a-5}{M}$, for a

2. ________________

3. $\frac{p}{w}=\frac{r}{p}$, for p

3. ________________

4. $\frac{D}{K}=N$, for K

4. ________________

5. $m - g = \frac{h}{m}$, for m

5. ________________

6. $A = P(1 + rt)$, for t

6. ________________

7. $I = \frac{nE}{E + nr}$, for r

7. ________________

8. $S = \frac{(v_1 + v_2)t}{2}$, for v_2

8. ________________

9. $S = \frac{H}{m(t_1 + t_2)}$, for t_2

9. ________________

Name: Date:
Instructor: Section:

Chapter 5 RATIONAL EXPRESSIONS, EQUATIONS, AND FUNCTIONS

5.8 Variation and Applications

Learning Objectives

a Find an equation of direct variation given a pair of values of the variables.
b Solve applied problems involving direct variation.
c Find an equation of inverse variation given a pair of values of the variables.
d Solve applied problems involving inverse variation.
e Find equations of other kinds of variation given values of the variables.
f Solve applied problems involving other kinds of variation.

Key Terms

Use the vocabulary terms listed below to complete each statement in Exercises 1–4.

direct **inverse** **proportionality** **variation**

1. The equation $y = kx$ is called an equation of ________________ variation.

2. In the equation $y = kx$, k is called the constant of ________________.

3. The equation $y = k/x$ is called an equation of ________________ variation.

4. In the equation $y = k/x$, k is called the ________________ constant.

Objective a Find an equation of direct variation given a pair of values of the variables.

Find the variation constant and an equation of variation in which y varies directly as x and the following are true.

5. $y = 55$ when $x = 5$ **5.** ________________

6. $y = \frac{2}{3}$ when $x = 8$ **6.** ________________

Objective b Solve applied problems involving direct variation.

Solve.

7. Hooke's law states that the distance d that a spring is stretched by a hanging object varies directly as the mass m of the object. If the distance is 50 cm when the mass is 6 kg, what is the distance when the mass is 2 kg?

7. ______________

8. The electric current I, in amperes, in a circuit varies directly as the voltage V. When 12 volts are applied, the current is 3 amperes. What is the current when 16 volts are applied?

8. ______________

9. The number of calories burned by a person in a Zumba aerobic class is directly proportional to the time spent exercising. It takes 10 min to burn 80 calories (*Source*: Family Fun and Fitness). How long would it take to burn 200 calories in the class?

9. ______________

Objective c Find an equation of inverse variation given a pair of values of the variables.

Find the variation constant and an equation of variation in which y varies inversely as x and the following are true.

10. $y = 7$ when $x = 9$

10. ______________

Name: Date:
Instructor: Section:

11. $y = 72$ when $x = \frac{1}{12}$ **11.** ____________

Objective d Solve applied problems involving inverse variation.

Solve.

12. The time T required to do a job varies inversely as the number of people P working. It takes 4 hr for 9 people to weed the community garden. How long would it take 10 people to complete the job? **12.** ____________

13. The wavelength W of a radio wave varies inversely as its frequency F. A wave with a frequency of 1600 kilohertz has a length of 225 meters. What is the length of a wave with a frequency of 3000 kilohertz? **13.** ____________

14. The current I in an electrical conductor varies inversely as the resistance R in the conductor. If the current is $\frac{2}{5}$ ampere when the resistance is 120 ohms, what is the current when the resistance is 150 ohms? **14.** ____________

Objective e Find equations of other kinds of variation given values of the variables.

Find an equation of variation in which the following are true.

15. y varies inversely as the square of x, and $y = 72$ when $x = 0.5$. **15.** ____________

16. y varies directly as the square of x, and $y = 0.44$ when $x = 0.1$.

16. ______________

17. y varies jointly as x and the square of z, and $y = 385$ when $x = 14$ and $z = 5$.

17. ______________

18. y varies jointly as w and the square of x and inversely as z, and $y = 9$ when $w = 3, x = 3$ and $z = 15$.

18. ______________

Objective f Solve applied problems involving other kinds of variation.

Solve.

19. The intensity I of a wireless signal varies inversely as the square of the distance d from the transmitter. If the intensity is 50 W/m^2 at a distance of 5 km, what is the intensity 20 km from the transmitter?

19. ______________

20. Atmospheric drag varies jointly as an object's surface area A and its velocity v. If a car traveling at a velocity of 80 mph with a surface area of 41.2 sq ft experiences a drag of 170 N, how fast must a car with 42.8 sq ft of surface area go to experience 409 N of drag? Round to the nearest integer.

20. ______________

Name: Date:
Instructor: Section:

Chapter 6 RADICAL EXPRESSIONS, EQUATIONS, AND FUNCTIONS

6.1 Radical Expressions and Functions

Learning Objectives

a Find principal square roots and their opposites, approximate square roots, identify radicands, find outputs of square-root functions, graph square-root functions, and find the domains of square-root functions.
b Simplify radical expressions with perfect-square radicands.
c Find cube roots, simplifying certain expressions, and find outputs of cube-root functions.
d Simplify expressions involving odd roots and even roots.

Key Terms

Use the vocabulary terms listed below to complete each statement in Exercises 1–6.

cube	**index**	**irrational**
principal	**radicand**	**square**

1. The number c is a(n) ____________________ root of a if $c^2 = a$.

2. The ____________________ square root of a nonnegative number is its nonnegative square root.

3. The square root of any whole number that is not a perfect square is a(n) ____________________ number.

4. The ____________________ is the expression under the radical sign.

5. The number c is a(n) ____________________ root of a if $c^3 = a$.

6. The number n in the expression $\sqrt[n]{a}$ is called the ____________________.

Objective a Find principal square roots and their opposites, approximate square roots, identify radicands, find outputs of square-root functions, graph square-root functions, and find the domains of square-root functions.

Find the square roots.

7. 64 7. ____________________

8. 4900   8. ______________

Simplify.

9. $-\sqrt{\dfrac{225}{169}}$ 9. ______________

10. $\sqrt{256}$ 10. ______________

11. $\sqrt{0.09}$ 11. ______________

12. $-\sqrt{0.0121}$ 12. ______________

Use a calculator to approximate to three decimal places.

13. $\sqrt{562}$ 13. ______________

14. $\sqrt{\dfrac{893}{147}}$ 14. ______________

15. $-\sqrt{762.8}$ 15. ______________

Name: Date:
Instructor: Section:

Identify the radicand.

16. $7\sqrt{x+1}+5$ **16.** ____________

17. $s\sqrt[3]{\dfrac{2s}{3t}}$ **17.** ____________

For the given function, find the indicated function values.

18. $f(x)=\sqrt{3x-12};\ f(5),\ f(4),\ f(0),\ f(-5)$ **18.** ____________

19. $g(x)=\sqrt{10-x^2};\ g(5),\ g(3),\ g(0),\ g(-2)$ **19.** ____________

20. Find the domain of the function $g(x)=\sqrt{x-4}$. **20.** ____________

Graph.

21. $f(x) = \sqrt{x+5}$

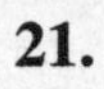

21.

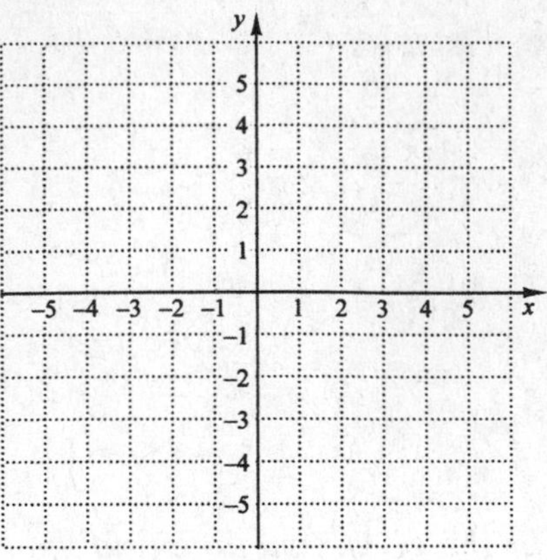

22. $g(x) = \sqrt{x} + 5$

22.

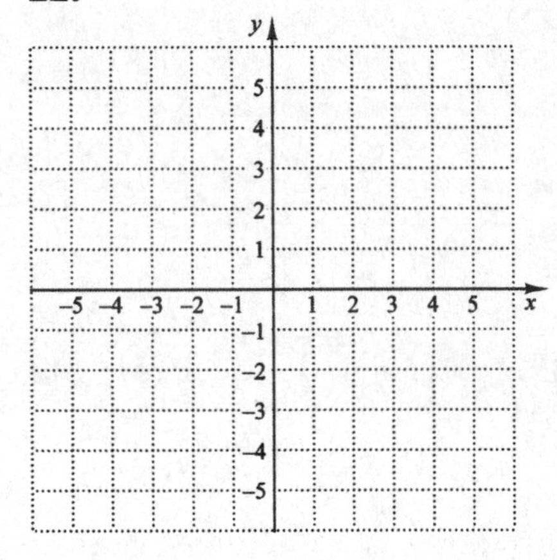

23. $g(x) = \sqrt{x} - 2$

23.

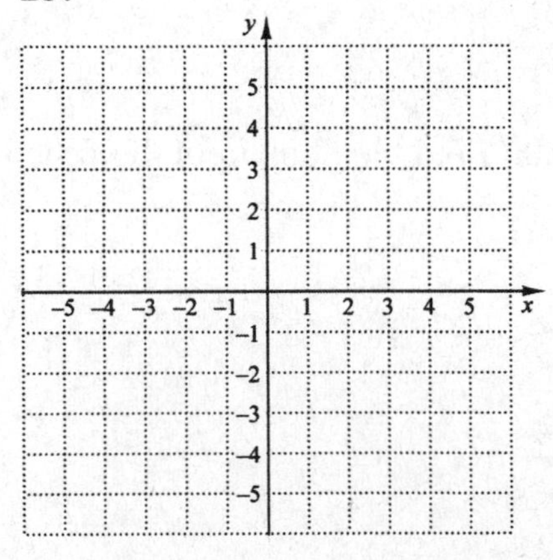

Name: Date:
Instructor: Section:

Objective b Simplify radical expressions with perfect-square radicands.

Find each of the following. Assume that letters can represent any *real number.*

24. $\sqrt{100m^2}$ **24.** ______________

25. $\sqrt{y^2-10y+25}$ **25.** ______________

26. $\sqrt{81c^2}$ **26.** ______________

27. $\sqrt{4x^2-28x+49}$ **27.** ______________

Objective c Find cube roots, simplifying certain expressions, and find outputs of cube-root functions.

Simplify.

28. $\sqrt[3]{-8}$ **28.** ______________

29. $\sqrt[3]{-125t^3}$ **29.** ______________

For the given function, find the indicated function values.

30. $g(x) = \sqrt[3]{2x-1};\ g(0),\ g(14),\ g(-13),\ g(-62)$ **30.** ______________

31. $f(t) = \sqrt[4]{t-10};\ f(9),\ f(11),\ f(-6),\ f(26)$ **31.** ______________

Objective d Simplify expressions involving odd roots and even roots.

Find each of the following. Assume that letters can represent any *real number.*

32. $\sqrt[6]{(5x)^6}$ **32.** ______________

33. $\sqrt[3]{-\frac{8}{27}}$ **33.** ______________

34. $\sqrt[4]{10{,}000}$ **34.** ______________

35. $\sqrt[9]{(3ab)^9}$ **35.** ______________

Name: Date:
Instructor: Section:

Chapter 6 RADICAL EXPRESSIONS, EQUATIONS, AND FUNCTIONS

6.2 Rational Numbers as Exponents

Learning Objective
a Write expressions with or without rational exponents, and simplify, if possible.
b Write expressions without negative exponents, and simplify, if possible.
c Use the laws of exponents with rational exponents.
d Use rational exponents to simplify radical expressions.

Key Terms
In Exercises 1–6, match each expression with an equivalent expression in the right column.

1. _____ a^{-n}

2. _____ $a^{m/n}$

3. _____ $a^m \cdot a^n$

4. _____ $\dfrac{a^m}{a^n}$

5. _____ $(a^m)^n$

6. _____ $(ab)^m$

a) a^{m-n}

b) $\sqrt[n]{a^m}$

c) $\dfrac{1}{a^n}$

d) a^{m+n}

e) $a^m b^m$

f) a^{mn}

Objective a Write expressions with or without rational exponents, and simplify, if possible.

Rewrite without rational exponents, and simplify, if possible.

7. $a^{1/4}$

7. ________________

8. $1000^{1/3}$

8. ________________

9. $(pqr)^{1/2}$ 9. ______________

10. $(x^3y^4)^{1/5}$ 10. ______________

11. $8^{5/3}$ 11. ______________

12. $(36c^6)^{3/2}$ 12. ______________

Rewrite without rational exponents.

13. $\sqrt[5]{10}$ 13. ______________

14. $\sqrt{x^7}$ 14. ______________

15. $\sqrt[4]{ab^3}$ 15. ______________

16. $(\sqrt[5]{3mn^4})^2$ 16. ______________

Name: Date:
Instructor: Section:

Objective b Write expressions without negative exponents, and simplify, if possible.

Rewrite with positive exponents, and simplify, if possible.

17. $(3x)^{-1/5}$ **17.** ______________

18. $\left(\frac{1}{27}\right)^{-2/3}$ **18.** ______________

19. $\frac{3x^2}{y^{-4/5}}$ **19.** ______________

20. $9^{-1/2}a^{2/3}b^{-3/5}$ **20.** ______________

Objective c Use the laws of exponents with rational exponents.

Use the laws of exponents to simplify. Write the answers with positive exponents.

21. $3^{2/5} \cdot 3^{1/10}$ **21.** ______________

22. $\frac{5^{-1/3}}{5^{-3/4}}$ **22.** ______________

23. $\left(x^{4/5}\right)^{3/2}$ **23.** ______________

24. $\left(a^{2/5}b^{-1/5}\right)^{-1/2}$ **24.** ______________

Objective d Use rational exponents to simplify radical expressions.

Use rational exponents to simplify. Write the answer in radical notation if appropriate.

25. $\sqrt[3]{x^{15}}$ **25.** ____________

26. $\left(\sqrt[6]{cd}\right)^3$ **26.** ____________

27. $\sqrt{\sqrt[5]{m}}$ **27.** ____________

28. $\left(\sqrt[4]{x^3y^2}\right)^{12}$ **28.** ____________

Use rational exponents to write a single radical expression.

29. $\sqrt[3]{6}\sqrt{6}$ **29.** ____________

30. $\sqrt[5]{\sqrt[3]{x}}$ **30.** ____________

31. $\left(\sqrt[4]{a^3b^2}\right)^{12}$ **31.** ____________

32. $\dfrac{x^{7/12} \cdot y^{5/12}}{x^{1/3} \cdot y^{-1/2}}$ **32.** ____________

Name: Date:
Instructor: Section:

Chapter 6 RADICAL EXPRESSIONS, EQUATIONS, AND FUNCTIONS

6.3 Simplifying Radical Expressions

Learning Objectives
a Multiply and simplify radical expressions.
b Divide and simplify radical expressions.

Key Terms
Use the vocabulary terms listed below to complete each statement in Exercises 1–4.

index **multiples** **radicands** **square**

1. The product of two *n*th roots is the *n*th root of the product of the two

____________________.

2. The product rule for radicals applies only when radicals have the same

____________________.

3. When no index is written, roots are understood to be ____________________ roots.

4. To simplify an *n*th root, identify factors in the radicand with exponents that are

____________________ of *n*.

Objective a Multiply and simplify radical expressions.

Simplify by factoring. Assume that no radicands were formed by raising negative numbers to even powers.

5. $\sqrt{200}$ **5.** ____________________

6. $\sqrt{63}$ **6.** ____________________

7. $\sqrt[3]{27x^6y^2}$ 7. ______________

8. $-\sqrt[3]{24x^{12}}$ 8. ______________

9. $\sqrt{x^5y^{10}}$ 9. ______________

10. $\sqrt[3]{a^6b^7c^{14}}$ 10. ______________

11. $\sqrt[5]{-320c^8d^{17}}$ 11. ______________

12. $\sqrt[4]{32x^3y^{20}z^{11}}$ 12. ______________

Multiply and simplify. Assume that no radicands were formed by raising negative numbers to even powers.

13. $\sqrt{10}\sqrt{11}$ 13. ______________

14. $\sqrt[5]{3x^2}\sqrt[5]{5x}$ 14. ______________

15. $\sqrt{14}\sqrt{21}$ 15. ______________

16. $\sqrt[3]{5}\sqrt[3]{25}$ 16. ______________

17. $\sqrt{28x^5}\sqrt{28x^5}$ 17. ______________

Name: Date:
Instructor: Section:

18. $\sqrt[3]{4a^2}\sqrt[3]{6a^5}$

18. ______________

19. $\sqrt[3]{c^4d^8}\sqrt[3]{cd^2}$

19. ______________

20. $\sqrt[4]{50x^5y^3}\sqrt[4]{75x^2y}$

20. ______________

Objective b Divide and simplify radical expressions.

Divide and simplify. Assume that all expressions under radicals represent positive numbers.

21. $\dfrac{\sqrt{30x}}{\sqrt{5x}}$

21. ______________

22. $\dfrac{\sqrt[3]{30}}{\sqrt[3]{3}}$

22. ______________

23. $\dfrac{\sqrt{6a^5b}}{\sqrt{2a}}$

23. ______________

24. $\dfrac{\sqrt{81xyz}}{3\sqrt{3}}$

24. ______________

Simplify.

25. $\sqrt{\frac{81}{49}}$

25. ______________

26. $\sqrt[3]{\frac{125}{64}}$

26. ______________

27. $\sqrt{\frac{16p^5}{q^{10}}}$

27. ______________

28. $\sqrt[3]{\frac{1000x^7}{27y^3}}$

28. ______________

29. $\sqrt[4]{\frac{a^8b^7}{c^{14}}}$

29. ______________

30. $\sqrt[6]{\frac{m^7n^6}{p^{20}}}$

30. ______________

Name: Date:
Instructor: Section:

Chapter 6 RADICAL EXPRESSIONS, EQUATIONS, AND FUNCTIONS

6.4 Addition, Subtraction, and More Multiplication

Learning Objectives

a Add or subtract with radical notation and simplify.

b Multiply expressions involving radicals in which some factors contain more than one term.

Key Terms

Use the vocabulary terms listed below to complete each statement in Exercises 1–2.

conjugates **like radicals**

1. When two radical expressions have the same indices and radicands, they are called

________________.

2. Pairs of radical terms like $\sqrt{a}+\sqrt{b}$ and $\sqrt{a}-\sqrt{b}$ are called ________________.

Objective a Add or subtract with radical notation and simplify.

Add or subtract. Then simplify by collecting like radical terms, if possible. Assume that no radicands were formed by raising negative numbers to even powers.

3. $5\sqrt{11}-8\sqrt{11}$ **3.** ____________

4. $\sqrt[5]{x}+15\sqrt[5]{x}$ **4.** ____________

5. $4\sqrt{3}+2\sqrt[3]{5}-\sqrt{3}+8\sqrt[3]{5}$ **5.** ____________

6. $6\sqrt{18}-5\sqrt{8}$ **6.** ____________

7. $\sqrt[3]{40a^7}+9\sqrt[3]{5a^4}$ 7. ______________

8. $3\sqrt{98}-5\sqrt{50}$ 8. ______________

9. $\sqrt{16z-4}-\sqrt{4z^3-z^2}$ 9. ______________

Objective b Multiply expressions involving radicals in which some factors contain more than one term.

Multiply. Assume that no radicands were formed by raising negative numbers to even powers.

10. $2\sqrt{3}\left(\sqrt{3}-\sqrt{10}\right)$ 10. ______________

11. $\sqrt[3]{4}\left(3\sqrt[3]{2}+5\sqrt[3]{10}\right)$ 11. ______________

12. $\left(3+\sqrt{5}\right)\left(7-\sqrt{5}\right)$ 12. ______________

13. $\left(2\sqrt{5}+3\sqrt{2}\right)\left(3\sqrt{5}-6\sqrt{2}\right)$ 13. ______________

14. $\left(6+\sqrt{3}\right)\left(6-\sqrt{3}\right)$ 14. ______________

15. $\left(\sqrt{2x}-\sqrt{7}\right)^2$ 15. ______________

16. $\left(\sqrt[3]{9}-\sqrt[3]{3}\right)\left(\sqrt[3]{2}+\sqrt[3]{5}\right)$ 16. ______________

17. $\left(\sqrt[4]{9}-\sqrt[4]{3}\right)\left(\sqrt[4]{8}+\sqrt[4]{25}\right)$ 17. ______________

Name: Date:
Instructor: Section:

Chapter 6 RADICAL EXPRESSIONS, EQUATIONS, AND FUNCTIONS

6.5 More on Division of Radical Expressions

Learning Objective

a Rationalize the denominator of a radical expression having one term in the denominator.

b Rationalize the denominator of a radical expression having two terms in the denominator.

Objective a Rationalize the denominator of a radical expression having one term in the denominator.

Rationalize the denominator. Assume that no radicands were formed by raising negative numbers to even powers.

1. $\frac{3\sqrt{7}}{5\sqrt{2}}$

1. ________________

2. $\frac{\sqrt[3]{2s}}{\sqrt[3]{3t}}$

2. ________________

3. $\sqrt[3]{\frac{10}{xy^2}}$

3. ________________

4. $\sqrt{\frac{8}{50x^2y}}$

4. ________________

5. $\sqrt{\frac{15}{35}}$

5. ________________

6. $\dfrac{5\sqrt{11}}{4\sqrt{3}}$

6. ________________

7. $\dfrac{\sqrt[3]{5}}{\sqrt[3]{7}}$

7. ________________

8. $\sqrt[3]{\dfrac{6x^7}{5y}}$

8. ________________

Objective b Rationalize the denominator of a radical expression having two terms in the denominator.

Rationalize the denominator. Assume that no radicands were formed by raising negative numbers to even powers.

9. $\dfrac{2}{5+\sqrt{3}}$

9. ________________

10. $\dfrac{6+\sqrt{7}}{2-\sqrt{5}}$

10. ________________

11. $\dfrac{\sqrt{p}}{\sqrt{p}-\sqrt{q}}$

11. ________________

12. $\dfrac{2\sqrt{11}-4\sqrt{5}}{6\sqrt{2}+3\sqrt{5}}$

12. ________________

Name: Date:
Instructor: Section:

Chapter 6 RADICAL EXPRESSIONS, EQUATIONS, AND FUNCTIONS

6.6 Solving Radical Equations

Learning Objective
a Solve radical equations with one radical term.
b Solve radical equations with two radical terms.
c Solve applied problems involving radical equations.

Key Terms
Use the vocabulary terms listed below to complete each statement in Exercises 1 and 2.

principle of powers **radical equation**

1. A ____________________ is an equation in which the variable appears in a radicand.

2. The ____________________ states that if $a = b$, then $a^n = b^n$ for any exponent n.

Objective a Solve radical equations with one radical term.

Solve.

3. $\sqrt{3x-7} = 5$ **3.** ________________

4. $\sqrt{6x} - 3 = 7$ **4.** ________________

5. $\sqrt{n-8} - 2 = 1$ **5.** ________________

6. $\sqrt{z+2} + 6 = 10$ **6.** ________________

7. $\sqrt[3]{t+3}=4$ 7. ______________

8. $\sqrt[4]{y-9}=2$ 8. ______________

9. $7\sqrt{a}=a$ 9. ______________

10. $6\sqrt{x}-10=2$ 10. ______________

11. $\sqrt[3]{x}=-10$ 11. ______________

12. $\sqrt{n+5}=-6$ 12. ______________

Objective b Solve radical equations with two radical terms.

Solve.

13. $\sqrt{6z+1}=\sqrt{5z+3}$ 13. ______________

14. $7+2\sqrt{x+1}=x$ 14. ______________

Name: Date:
Instructor: Section:

15. $\sqrt{5x+1}=1+\sqrt{4x-3}$ **15.** ____________

16. $7+\sqrt{10-z}=8+\sqrt{1-z}$ **16.** ____________

17. $2\sqrt{3y-2}-\sqrt{4y-3}=1$ **17.** ____________

Objective c Solve applied problems involving radical equations.

Solve.

18. How far can you see to the horizon from a given height? The function $D=1.2\sqrt{h}$ can be used to approximate the distance D, in miles, that a person can see to the horizon from a height h, in feet.
A person can see 210 mi to the horizon from an airplane window. How high is the airplane?

18. ____________

19. How far can you see to the horizon from a given height? The function $D=1.2\sqrt{h}$ can be used to approximate the distance D, in miles, that a person can see to the horizon from a height h, in feet.
Maddie can see 29.2 mi to the horizon from the top of a cliff. What is the height of his eyes?

19. ____________

20. How far can you see to the horizon from a given height? The function $D = 1.2\sqrt{h}$ can be used to approximate the distance D, in miles, that a person can see to the horizon from a height h, in feet.
A technician can see 35.2 mi to the horizon from the top of a radio tower. How high is the tower?

20. ______________

21. The formula $r = 2\sqrt{5L}$ can be used to approximate the speed r, in miles per hour, of a car that has left a skid mark of length L, in feet.
How far will a car skid at 60 mph?

21. ______________

22. The formula $r = 2\sqrt{5L}$ can be used to approximate the speed r, in miles per hour, of a car that has left a skid mark of length L, in feet.
How far will a car skid at 70 mph?

22. ______________

23. The formula $T = 2\pi\sqrt{L/32}$ can be used to find the period T, in seconds, of a pendulum of length L, in feet. The pendulum in a grandfather clock has a period of 2.1 sec. Find the length of the pendulum. Use 3.14 for π.

23. ______________

24. The formula $T = 2\pi\sqrt{L/32}$ can be used to find the period T, in seconds, of a pendulum of length L, in feet. A playground swing has a period of 3.2 sec. Find the length of the swing's chain. Use 3.14 for π.

24. ______________

Name: Date:
Instructor: Section:

Chapter 6 RADICAL EXPRESSIONS, EQUATIONS, AND FUNCTIONS

6.7 Applications Involving Powers and Roots

Learning Objectives

a Solve applied problems involving the Pythagorean theorem and powers and roots.

Objective a Solve applied problems involving the Pythagorean theorem and powers and roots.

In a right triangle, find the length of a side not given. Give an exact answer and, where appropriate, an approximation to three decimal places.

1. $a=4,\ b=7$ **1.** ______________

2. $a=7,\ b=11$ **2.** ______________

3. $b=4,\ c=9$ **3.** ______________

4. $c=12,\ a=\sqrt{7}$ **4.** ______________

5. $b=2,\ c=\sqrt{26}$ **5.** ______________

6. $b=3,\ a=\sqrt{n}$ **6.** ______________

In the following problems, give an exact answer and, where appropriate, an approximation to three decimal places.

7. How long is a guy wire reaching from the top of a 12-ft pole to a point on the ground 6 ft from the pole?

7. 

8. Using the formula $L = \dfrac{0.000169d^{2.27}}{h}$, find the length L of a road-pavement message when $h = 4$ ft and $d = 190$ ft.

8. 

9. A large screen television has a 40-inch diagonal with height 24 inches. What is its width?

9. 

10. The diagonal of a square has length $12\sqrt{2}$ ft. Find the length of a side of the square.

10. ______________

Name: Date:
Instructor: Section:

Chapter 6 RADICAL EXPRESSIONS, EQUATIONS, AND FUNCTIONS

6.8 The Complex Numbers

Learning Objective

a Express imaginary numbers as *bi*, where *b* is a nonzero real number, and complex numbers as *a* + *bi*, where *a* and *b* are real numbers.
b Add and subtract complex numbers.
c Multiply complex numbers.
d Write expressions involving powers of *i* in the form *a* + *bi*.
e Find conjugates of complex numbers and divide complex numbers.
f Determine whether a given complex number is a solution of an equation.

Key Terms

Use the vocabulary terms listed below to complete each statement in Exercises 1–4.

complex **conjugate** ***i*** **imaginary**

1. $\sqrt{-1} =$ ____________________.

2. A(n) ____________________ number is any number that can be written in the form $a+bi$, where a and b are real numbers and $b \neq 0$.

3. A(n) ____________________ number is any number that can be written in the form $a+bi$, where a and b are real numbers.

4. The ____________________ of a complex number $a+bi$ is $a-bi$.

Objective a Express imaginary numbers as *bi*, where *b* is a nonzero real number, and complex numbers as *a* + *bi*, where *a* and *b* are real numbers.

Express in terms of i.

5. $\sqrt{-49}$ **5.**________________

6. $\sqrt{-17}$

6.________________

7. $\sqrt{-200}$

7.________________

8. $-\sqrt{-45}$

8.________________

9. $8-\sqrt{-150}$

9.________________

10. $\sqrt{-32}-\sqrt{-9}$

10.________________

Objective b Add and subtract complex numbers.

Add or subtract and simplify.

11. $(3+5i)+(9-2i)$

11.________________

12. $(5-7i)+(11+4i)$

12.________________

13. $(3-8i)-(6-2i)$

13.________________

Name: Date:
Instructor: Section:

14. $(-2-5i)-(9-7i)$ **14.**________________

Objective c Multiply complex numbers.

Multiply.

15. $8i \cdot 11i$ **15.**________________

16. $\sqrt{-100} \cdot \sqrt{-81}$ **16.**________________

17. $2i(5-4i)$ **17.**________________

18. $-3i(6-i)$ **18.**________________

19. $(-2+3i)(5+6i)$ **19.**________________

20. $(7-3i)(3-4i)$

20.______________

21. $(4-7i)^2$

21.______________

22. $(10+3i)^2$

22.______________

Objective d Write expressions involving powers of *i* in the form *a* + *bi*.

Simplify.

23. i^{44}

23.______________

24. $(-i)^{10}$

24.______________

25. $(2i)^3$

25.______________

26. $(2i)^5$

26.______________

Name: Date:
Instructor: Section:

Objective e Find conjugates of complex numbers and divide complex numbers.

Divide and simplify to the form a + bi.

27. $\frac{3}{2+i}$

27.________________

28. $\frac{11}{5i}$

28.________________

29. $\frac{6-5i}{2i}$

29.________________

30. $\frac{3+2i}{5-3i}$

30.________________

Objective f Determine whether a given complex number is a solution of an equation.

Determine whether the complex number is a solution of the equation..

31. $1-3i$;
$x^2-2x+10=0$

31. ______________

32. $1+3i$;
$x^2-2x+10=0$

32. ______________

33. $3+4i$;
$x^2-6x+13=0$

33. ______________

Name: Date:
Instructor: Section:

Chapter 7 QUADRATIC EQUATIONS AND FUNCTIONS

7.1 The Basics of Solving Quadratic Equations

Learning Objectives

a Solve quadratic equations using the principle of square roots and find the x-intercepts of the graph of a related function.
b Solve quadratic equations by completing the square.
c Solve applied problems using quadratic equations.

Key Terms

Use the vocabulary terms listed below to complete each statement in Exercises 1–4.

complete **quadratic** **standard**

1. The equation $x^2 = 6 + x$ is an example of a(n) ________________ equation.

2. The ________________ form of a quadratic equation is $ax^2 + bx + c = 0$.

3. To ________________ the square for $x^2 + bx$, we add $\left(\frac{b}{2}\right)^2$.

Objective a Solve quadratic equations using the principle of square roots and find the x-intercepts of the graph of a related function.

4. a) Solve: $4x^2 = 24$.
b) Find the x-intercepts of $f(x) = x^2 - 24$.

4. a)________________

b)________________

5. a) Solve: $4x^2 = -81$.
b) Find the x-intercepts of $f(x) = 4x^2 - 81$.

5. a)________________

b)________________

Solve. Give the exact solution and approximate solutions to three decimal places, when appropriate.

6. $6t^2 = 18$ **6.** ______________

7. $4x^2 + 81 = 0$ **7.** ______________

8. $(y-3)^2 = 100$ **8.** ______________

9. $(x-5)^2 = 2$ **9.** ______________

10. $(a+7)^2 = -4$ **10.** ______________

11. $\left(y+\frac{2}{3}\right)^2 = \frac{8}{9}$ **11.** ______________

Objective b Solve quadratic equations by completing the square.

Solve by completing the square. Show your work.

12. $x^2 + x = 20$ **12.** ______________

13. $t^2 - 6t = -8$ **13.** ______________

Name: Date:
Instructor: Section:

14. $t^2 - 4t + 2 = 0$ **14.** ______________

15. $x^2 + 6x - 2 = 0$ **15.** ______________

Complete the square to find the x-intercepts of each function.

16. $f(x) = x^2 + 8x + 9$ **16.** ______________

17. $g(x) = x^2 - 4x + 2$ **17.** ______________

18. $f(x) = 2x^2 + x - 5$ **18.** ______________

19. $f(x) = 4x^2 - 2x - 1$ **19.** ______________

Solve by completing the square. Show your work.

20. $8x^2 + 3 = 10x$ **20.** ______________

21. $3x^2 + 2x - 4 = 0$ **21.** ______________

22. $x^2 + x + 8 = 0$ **22.** ______________

23. $x^2 + 8x + 25 = 0$ **23.** ______________

Objective c Solve applied problems using quadratic equations.

24. LeBron James has a vertical leap of 44 in. (*Source*: HoopsVibe.com). Use $V = 48T^2$ to estimate his hang time. **24.** ______________

The function $s(t) = 16t^2$ *is used to approximate the distance s, in feet, that an object falls freely from rest in t seconds. Use the formula for Exercises 25 and 26.*

25. The Nurek Dam in Tajikistan, at 984 ft, is the world's tallest dam. How long would it take an object to fall freely from the top? **25.** ______________

26. The Spire of Dublin in Ireland, at 393 ft, is the world's tallest sculpture. How long would it take an object to fall freely from the top? **26.** ______________

Name: Date:
Instructor: Section:

Chapter 7 QUADRATIC EQUATIONS AND FUNCTIONS

7.2 The Quadratic Formula

Learning Objective

a Solve quadratic equations using the quadratic formula, and approximate solutions using a calculator.

Key Terms

In Exercises 1 and 2, complete each statement.

1. The principle of square roots states that for any real number *k*, if $x^2 = k$, then

$x =$ ____________________.

2. The quadratic formula states that the solutions of $ax^2 + bx + c = 0, a \neq 0$, are given by

$x =$ ____________________.

Objective a Solve quadratic equations using the quadratic formula, and approximate solutions using a calculator.

Solve.

3. $x^2 + 3x - 2 = 0$ **3.** ________________

4. $4a^2 = 32a - 20$ **4.** ________________

5. $x^2 + x + 3 = 0$ **5.** ________________

6. $t^2 + 11 = 4t$

6. ________________

7. $\frac{1}{x^2} - 2 = \frac{8}{x}$

7. ________________

8. $4s + s(s - 3) = 1$

8. ________________

9. $10x^2 + 5x = 2$

9. ________________

Name: Date:
Instructor: Section:

10. $100x^2 - 60x + 9 = 0$

10. ______________

11. $6x(x+1) + 3 = 5x(x+2)$

11. ______________

12. $5t^2 = 12t + 16$

12. ______________

13. $x^2 + 25 = 6x$

13. ______________

14. $x^3 - 125 = 0$

14. ______________

Solve. Give the exact solutions and approximate solutions to three decimal places.

15. $x^2 + 2x - 5 = 0$

15. ________________

16. $x^2 - 5x + 2 = 0$

16. ________________

17. $2x^2 + 4x - 7 = 0$

17. ________________

18. $3x^2 - 5x - 4 = 0$

18. ________________

Name: Date:
Instructor: Section:

Chapter 7 QUADRATIC EQUATIONS AND FUNCTIONS

7.3 Applications Involving Quadratic Equations

Learning Objectives

a Solve applied problems involving quadratic equations.

b Solve a formula for a given letter.

Objective a Solve applied problems involving quadratic equations.

Solve.

1. The width of a rectangular flower garden is 5 feet less than the length. The area is 36 ft^2. Find the length and width.

1. ______________

2. The length of a rectangular parking lot is three times the width. The area is 1728 ft^2. Find the length and the width.

2. ______________

3. The base of a triangular sail is 4 ft less than its height. The area is 72 ft^2. Find the base and the height of the sail.

3. ______________

4. The hypotenuse of a right triangle is 75 m long. The length of one leg is 51 m less than the other. Find the length of the legs.

4. ______________

5. A student opens a geology book to two facing pages. The product of the page numbers is 5852. Find the page numbers.

5. ______________

6. The outside of a mosaic mirror frame measures 21 in. by 30 in., and 360 in^2 of mirror shows. Find the width of the frame.

6. ______________

7. During the first part of a trip, Kai drove 150 mi at a certain speed. Kai then drove another 90 mi at a speed that was 5 mph slower. If the total time of Kai's trip was 5 hr, what was her speed on each part of the trip?

7. ______________

Name: Date:
Instructor: Section:

8. Daquan traveled 300 mi averaging a certain speed. If the car had gone 10 mph slower, the trip would have taken 1 hr longer. Find Daquan's average speed.

8. ______________

9. Tim flies 500 mi at a certain speed. David flies 1000 mi at a speed that is 75 mph faster, but takes 1 hr longer. Find the speed of each plane.

9. ______________

10. Hwae bikes the 48 mi to the state park averaging a certain speed. The return trip is made at a speed that is 2 mph slower. Total time for the round trip is 14 hr. Find Hwae's average speed on each part of the trip.

10. ______________

11. The White River flows at a rate of 4 mph. A boat travels 20 mi upriver and returns in a total time of 3 hr. What is the speed of the boat in still water?

11. ______________

Objective b Solve a formula for a given letter.

Solve each formula for the given letter. Assume that all variables represent nonnegative numbers.

12. $T = 8v^2$, for v — **12.** ____________

13. $y = 3x^2 + 5x$, for x — **13.** ____________

14. $y = \dfrac{kxz}{w^2}$, for w — **14.** ____________

15. $M = 5\sqrt{\dfrac{p}{q}}$, for q — **15.** ____________

16. $a^2 = b^2 + c^2 - 2abX$, for c — **16.** ____________

17. $t = x_0 v + \dfrac{av^2}{8}$, for v — **17.** ____________

18. $B = \frac{1}{3}(x^2 - 2x)$, for x — **18.** ____________

19. $H = 1.6\sqrt{d}$, for d — **19.** ____________

Name: Date:
Instructor: Section:

Chapter 7 QUADRATIC EQUATIONS AND FUNCTIONS

7.4 More on Quadratic Equations

Learning Objectives
a Determine the nature of the solutions of a quadratic equation.
b Write a quadratic equation having two given numbers as solutions.
c Solve equations that are quadratic in form.

Key Terms
Use the vocabulary terms listed below to complete each statement in Exercises 1–4.

conjugates **discriminant** **real**

1. The expression b^2-4ac in the quadratic formula is called the ________________.

2. When $b^2-4ac=0$ there is one ________________ number solution.

3. When b^2-4ac is negative, the solutions are complex ________________.

Objective a Determine the nature of the solutions of a quadratic equation.

Determine the nature of the solutions of each equation.

4. $x^2-3x+10=0$ **4.** ________________

5. $x^2-12=0$ **5.** ________________

6. $t^2+15=0$ **6.** ________________

7. $6x^2-5x=0$ **7.** ________________

8. $6x^2-19x+10=0$ **8.** ________________

9. $a^2 + a + 4 = 0$

9. ________________

10. $16x^2 + 24x + 9 = 0$

10. ________________

11. $3x^2 - 7x = -2$

11. ________________

Objective b Write a quadratic equation having two given numbers as solutions.

Write a quadratic equation having the given numbers as solutions.

12. $-5, 6$

12. ________________

13. 4, only solution

13. ________________

14. $-2, -10$

14. ________________

15. $4, -\frac{3}{5}$

15. ________________

16. $-\sqrt{5}, \sqrt{5}$

16. ________________

17. $3\sqrt{10}, -3\sqrt{10}$

17. ________________

18. $5i, -5i$

18. ________________

Name: Date:
Instructor: Section:

Objective c Solve equations that are quadratic in form.

Solve.

19. $x^4 - 17x^2 + 16 = 0$

19. ______________

20. $a^4 - 17a^2 + 72 = 0$

20. ______________

21. $t - 3\sqrt{t} - 10 = 0$

21. ______________

22. $(m^2 - 5)^2 - 8(m^2 - 5) + 15 = 0$

22. ______________

23. $m^{-2}+m^{-1}-30=0$ **23.** ________________

24. $6x^{-2}-19x^{-1}+10=0$ **24.** ________________

25. $w^{2/5}-3w^{1/5}-4=0$ **25.** ________________

Find the x-intercepts of the graph of each function.

26. $f(x)=3x+13\sqrt{x}-10$ **26.** ________________

27. $f(x)=(x^2-2x)^2-11(x^2-2x)+24$ **27.** ________________

28. $f(x)=x^{2/3}-x^{1/3}-20$ **28.** ________________

Name: Date:
Instructor: Section:

Chapter 7 QUADRATIC EQUATIONS AND FUNCTIONS

7.5 Graphing $f(x) = a(x - h)^2 + k$

Learning Objective

a Graph quadratic functions of the type $f(x) = ax^2$ and then label the vertex and the line of symmetry.

b Graph quadratic functions of the type $f(x) = a(x - h)^2$ and then label the vertex and the line of symmetry.

c Graph quadratic functions of the type $f(x) = a(x - h)^2 + k$, finding the vertex, the line of symmetry, and the maximum or minimum function value, or y-value.

Key Terms

Use the vocabulary terms listed below to complete each statement in Exercises 1–4.

axis of symmetry **parabola** **translated** **vertex**

1. The graph of a quadratic equation is a(n) ________________.

2. The graph of a quadratic equation is symmetric with respect to its ________________.

3. The maximum or minimum value of a quadratic function occurs at the ________________ of its graph.

4. The graph of $f(x) = a(x-h)^2 + k$ looks like the graph of $f(x) = ax^2$ except that it is ________________ $|h|$ units horizontally and $|k|$ units vertically.

Objective a Graph quadratic functions of the type $f(x) = ax^2$ and then label the vertex and the line of symmetry.

Objective b Graph quadratic functions of the type $f(x) = a(x - h)^2$ and then label the vertex and the line of symmetry.

For each of the following, graph the function, label the vertex, and draw the axis of symmetry.

5. $f(x) = (x+2)^2$

5.

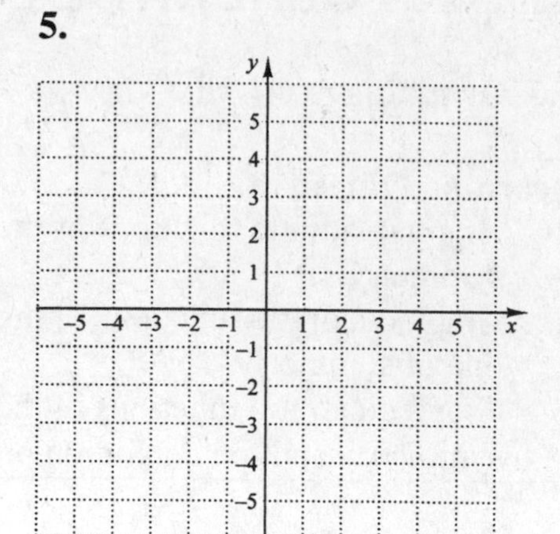

6. $f(x) = (x-6)^2$

6.

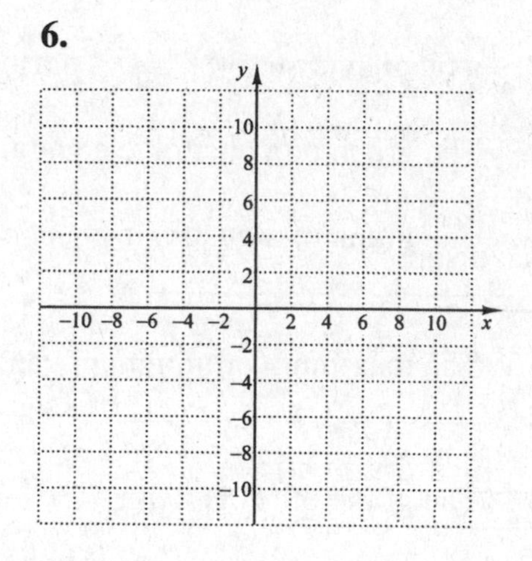

7. $g(x) = -(x-3)^2$

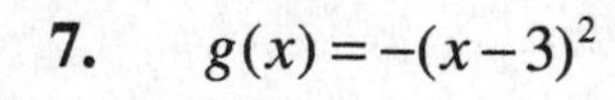

7.

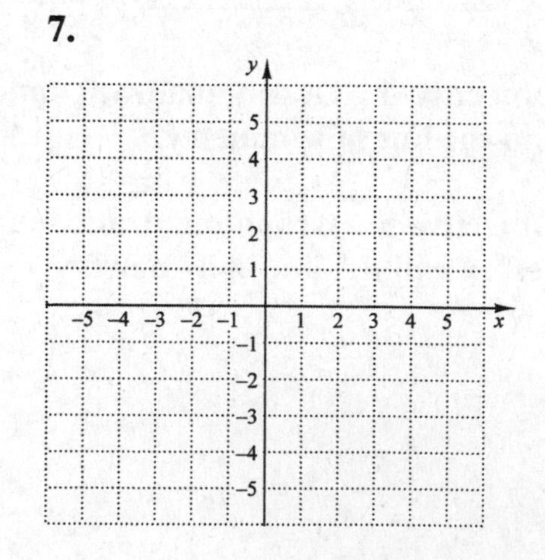

Name: Date:
Instructor: Section:

8. $g(x) = -2(x+1)^2$

8.

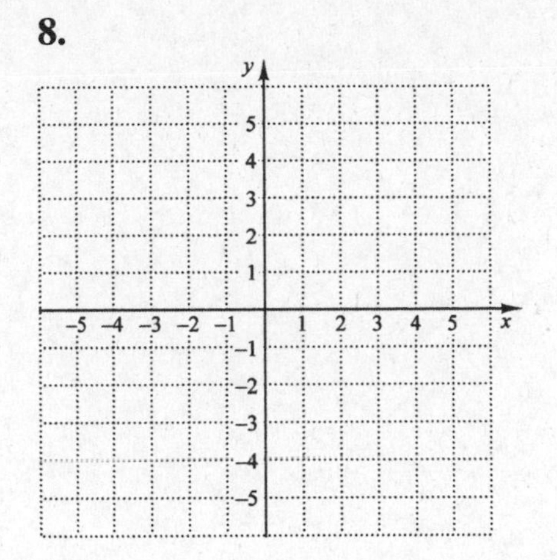

Objective c Graph quadratic functions of the type $f(x) = a(x-h)^2 + k$, finding the vertex, the line of symmetry, and the maximum or minimum function value, or y-value.

For each of the following, graph the function and find the vertex, the axis of symmetry, and the maximum value or the minimum value.

9. $f(x) = (x+2)^2 - 3$

9.________________

____

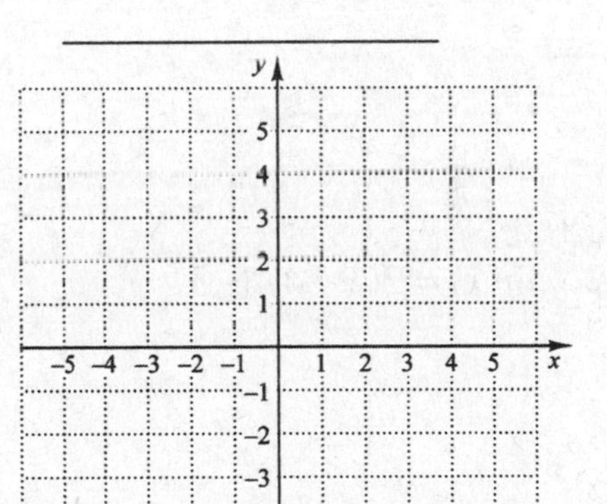

10. $f(x) = -(x-2)^2 + 1$

10.________________

____

11. $h(x) = -\frac{1}{2}(x-3)^2 + 4$

11.________________

____

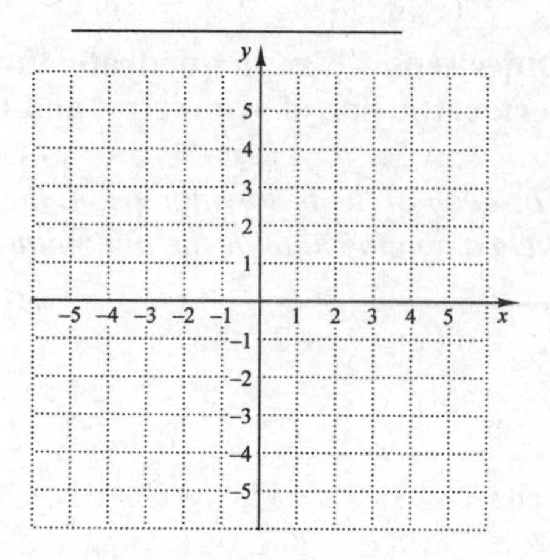

12. $h(x) = 2(x-4)^2 - 3$

12.________________

____

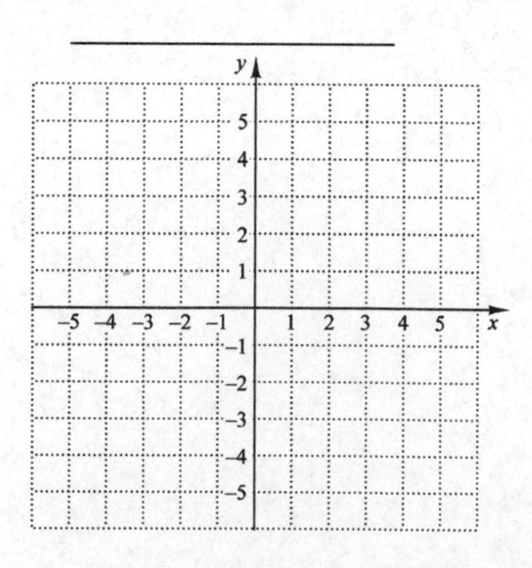

Name: Date:
Instructor: Section:

Chapter 7 QUADRATIC EQUATIONS AND FUNCTIONS

7.6 Graphing $f(x) = ax^2 + bx + c$

Learning Objective

a For a quadratic function, find the vertex, the line of symmetry, and the maximum or minimum value, and then graph the function.

b Find the intercepts of a quadratic function.

Key Terms

Use the vocabulary terms listed below to complete each statement in Exercises 1–4.

maximum **minimum** ***x*-intercept** ***y*-intercept**

1. The function given by $f(x) = a(x-h)^2 + k$ has a(n) ________________ value of k if a is positive.

2. The function given by $f(x) = a(x-h)^2 + k$ has a(n) ________________ value of k if a is negative.

3. For any function, the ________________ occurs at $f(0)$.

4. A(n) ________________ will occur at an input for which $f(x) = 0$.

Objective a For a quadratic function, find the vertex, the line of symmetry, and the maximum or minimum value, and then graph the function.

For each quadratic function, **a)** *write the function in the form* $f(x) = a(x-h)^2 + k$ *and* **b)** *find the vertex and the axis of symmetry.*

5. $f(x) = x^2 + 8x - 5$

5. a)____________

b)____________

For each quadratic function, find **a)** *the vertex and the axis of symmetry,* **b)** *the minimum or maximum value, and* **c)** *graph the function.*

6. $f(x) = x^2 + 2x + 3$

6. a)________________

b)________________

c)

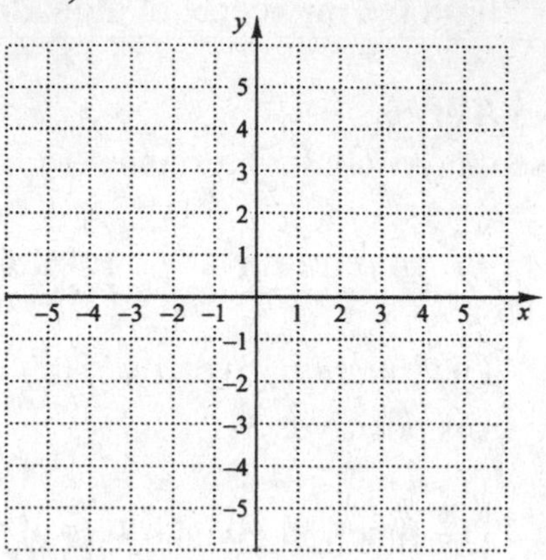

7. $f(x) = x^2 - 4x - 1$

7. a)________________

b)________________

c)

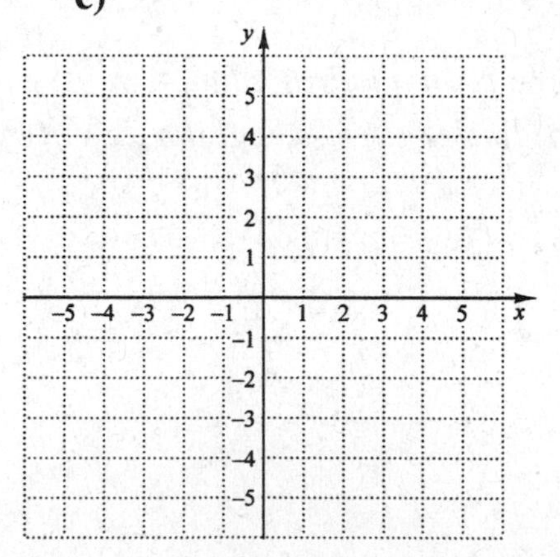

Name: Date:
Instructor: Section:

8. $h(x)=2x^2-8x+9$

8. a)________________

b)________________

c)

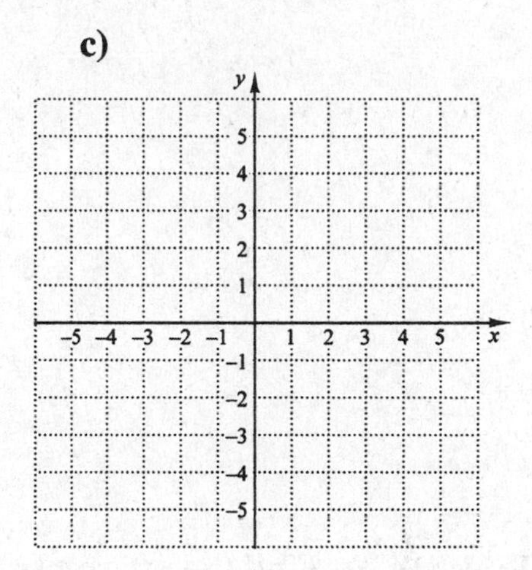

9. $f(x)=-x^2+2x-5$

9. a)________________

b)________________

c)

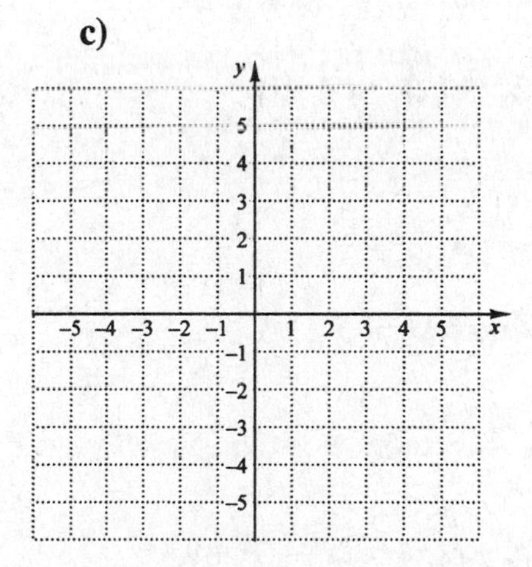

10. $f(x) = -\frac{1}{2}x^2 + 2x + 1$

10. a)________________

b)________________

c)

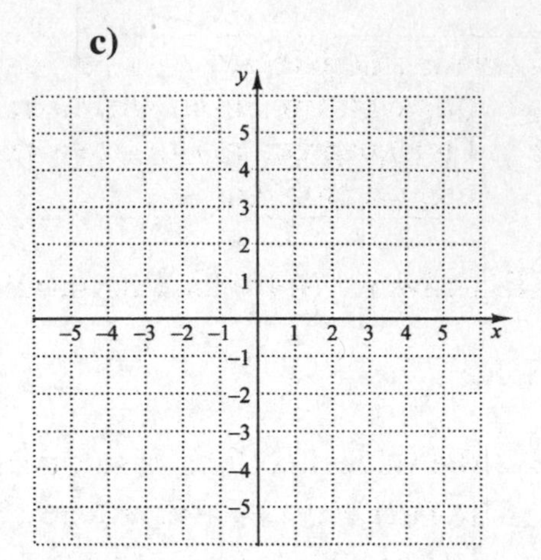

Objective b Find the intercepts of a quadratic function.

Find the x- and y-intercepts.

11. $f(x) = x^2 + 6x - 1$

11. ________________

12. $g(x) = x^2 - 10x$

12. ________________

13. $h(x) = -x^2 + x + 6$

13. ________________

14. $f(x) = 2x^2 - 3x + 8$

14. ________________

Name: Date:
Instructor: Section:

Chapter 7 QUADRATIC EQUATIONS AND FUNCTIONS

7.7 Mathematical Modeling with Quadratic Functions

Learning Objectives

a Solve maximum-minimum problems involving quadratic functions.

b Fit a quadratic function to a set of data to form a mathematical model, and solve related applied problems.

Objective a Solve maximum-minimum problems involving quadratic functions.

Solve.

1. The value of a share of stock can be represented by $f(x) = x^2 - 4x + 10$, where x is the number of months after January 2007. What is the lowest value $f(x)$ will reach, and when will that occur?

1. ______________

2. The average cost per box built at Green Box Company can be estimated by $c(x) = 0.1x^2 - 0.6x + 1.35$, where $c(x)$ is in dollars. What is minimum average cost per box and how many boxes should be built to achieve that minimum?

2. ______________

3. A rectangular garden has a perimeter of 84 ft. What dimensions yield the maximum area?

3. ______________

4. Nick is fencing in a rectangular section of his yard. He has 90 ft of fence, and will use a wall of his house as one side of the rectangle. What is the maximum area Nick can enclose? What should the dimensions of the rectangle be in order to yield this area?

4. ______________

5. A raised rectangular garden is formed in a corner of a fenced yard, with 10 ft of treated lumber completing the other two sides of the rectangle. If the lumber is 8 in. high, what dimensions of the base will maximize the garden's volume?

5. ______________

6. What is the maximum product of two numbers that add to 24? What are the numbers?

6. ______________

7. What is the minimum product of two numbers that differ by 10? What are the numbers?

7. ______________

Name: Date:
Instructor: Section:

8. What is the maximum product of two numbers that add to -14? What numbers yield this product?

8. ______________

Objective b Fit a quadratic function to a set of data to form a mathematical model, and solve related applied problems.

For the scatterplots and graphs in Exercises 9–12, determine which, if any, of the following functions might be used as a model for the data: Linear, $f(x)=mx+b$ *; quadratic,* $f(x)=ax^2+bx+c, a>0$ *; quadratic,* $f(x)=ax^2+bx+c, a<0$ *; polynomial, neither quadratic nor linear.*

9.

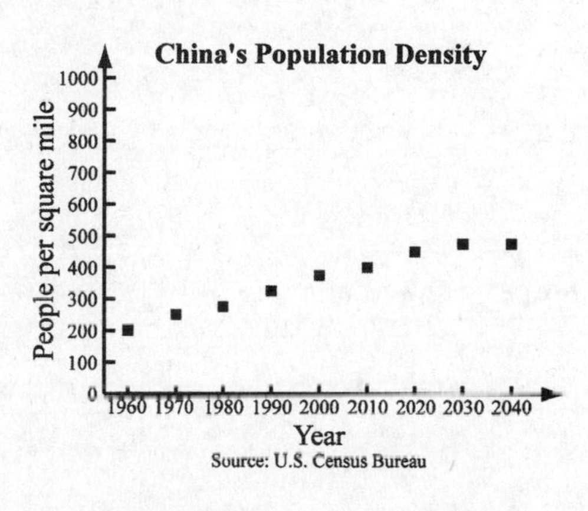

9. ______________

10.

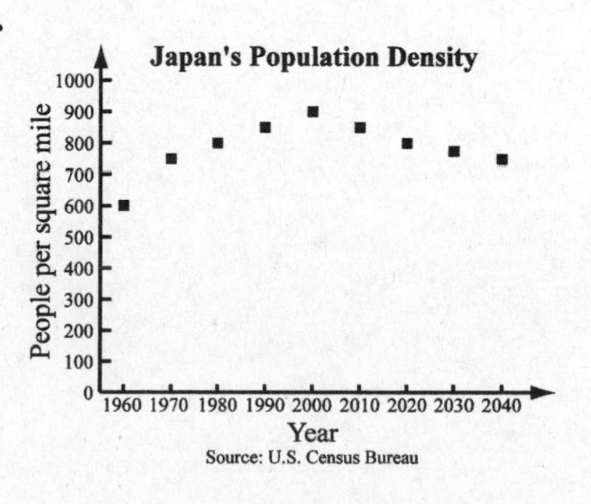

10. ______________

11.

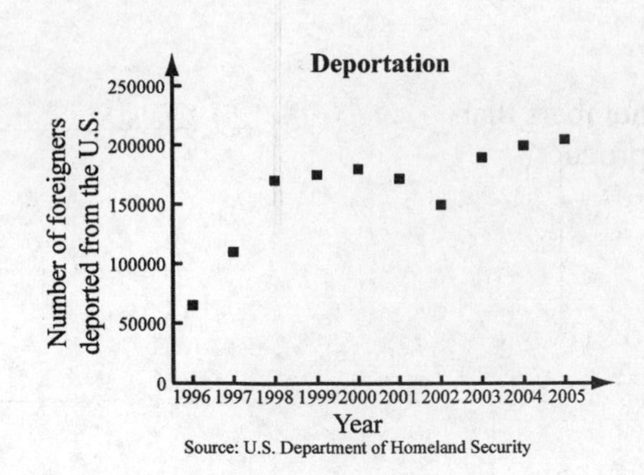

11. ______________

12.

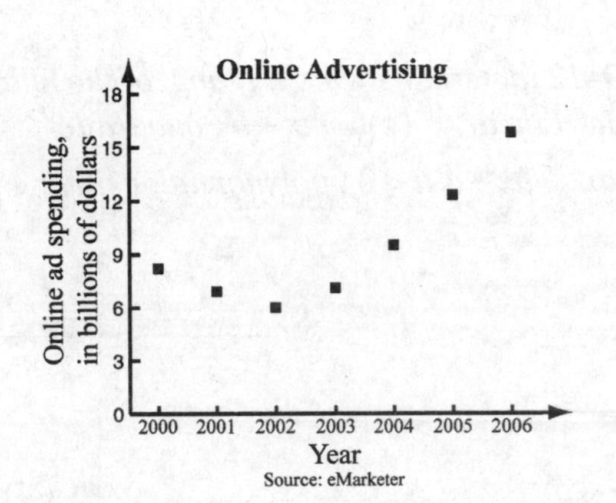

12. ______________

Find a quadratic function that fits the set of data points.

13. $(1,3),(-1,-1),(2,10)$

13. ______________

14. $(1,0),(-1,8),(2,6)$

14. ______________

Name: Date:
Instructor: Section:

Chapter 7 QUADRATIC EQUATIONS AND FUNCTIONS

7.8 Polynomial Inequalities and Rational Inequalities

Learning Objective
a Solve quadratic inequalities and other polynomial inequalities.
b Solve rational inequalities.

Key Terms
Use the vocabulary terms listed below to complete each statement in Exercises 1–4.

polynomial **rational** **test points** **zeros**

1. An inequality like $x^3 - x > 6x^2 + 7$ is an example of a _________________ inequality.

2. The ___________________ of a function occur at the x-intercepts of the graph of the function.

3. We use ___________________ to determine the sign of a polynomial over an interval of the x-axis.

4. An inequality like $\frac{x+4}{2x+1} \leq 0$ is an example of a _________________ inequality.

Objective a Solve quadratic inequalities and other polynomial inequalities.

Solve algebraically and verify results from the graph.

5. $(x+1)(2x-9) < 0$ **5.** _______________

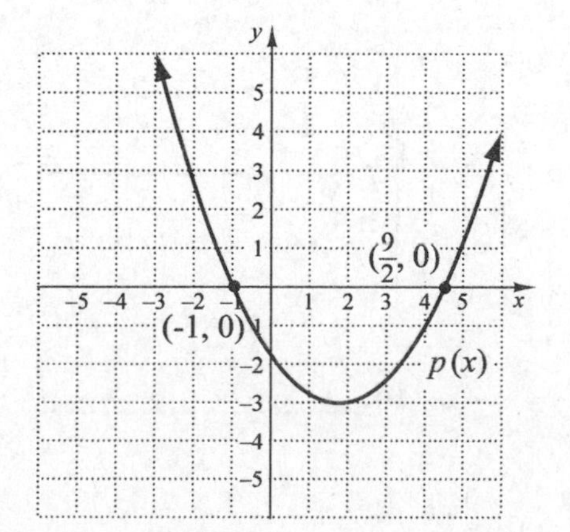

Solve.

6. $(x+2)(x-5)>0$ 6. ________

7. $(x-4)(x+1)\geq 0$ 7. ________

8. $x^2+x-6\leq 0$ 8. ________

9. $x^2+2x+1<0$ 9. ________

10. $5x(x-3)(x+6)<0$ 10. ________

Objective b Solve rational inequalities.

Solve.

11. $\dfrac{1}{x-3}>0$ 11. ________

12. $\dfrac{x+2}{x-1}\leq 4$ 12. ________

13. $\dfrac{(x+3)(x-5)}{x+2}\geq 0$ 13. ________

14. $\dfrac{x}{7-x}>0$ 14. ________

Name: Date:
Instructor: Section:

Chapter 8 EXPONENTIAL AND LOGARITHMIC FUNCTIONS

8.1 Exponential Functions

Learning Objectives
a Graph exponential equations and functions.
b Graph exponential equations in which x and y have been interchanged.
c Solve applied problems involving applications of exponential functions and their graphs.

Key Terms
Use the vocabulary terms listed below to complete each statement in Exercises 1–4.

asymptote **decreasing** **exponential** **increasing**

1. A function in the form $f(x)=a^x$ is a(n) ________________ function.

2. A function f is ________________ if the values of $f(x)$ increase as x increases.

3. A function f is ________________ if the values of $f(x)$ decrease as x increases.

4. The graph of a function $f(x)=a^x$ has the x-axis as a(n) ________________.

Objective a Graph exponential equations and functions.

Graph.

5. $f(x)=4^x$

5.

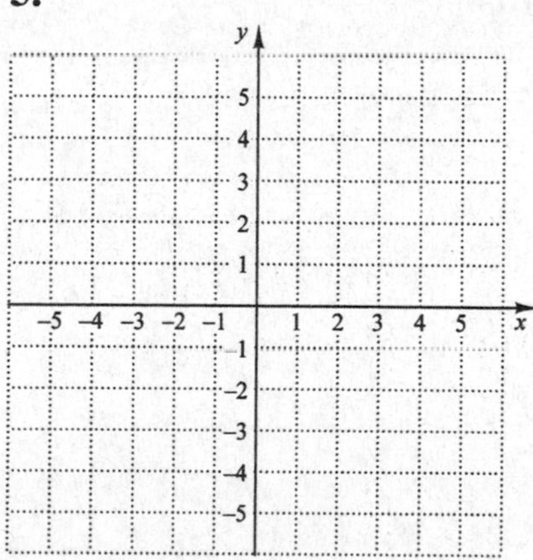

6. $f(x) = 2^{x-3}$

6.

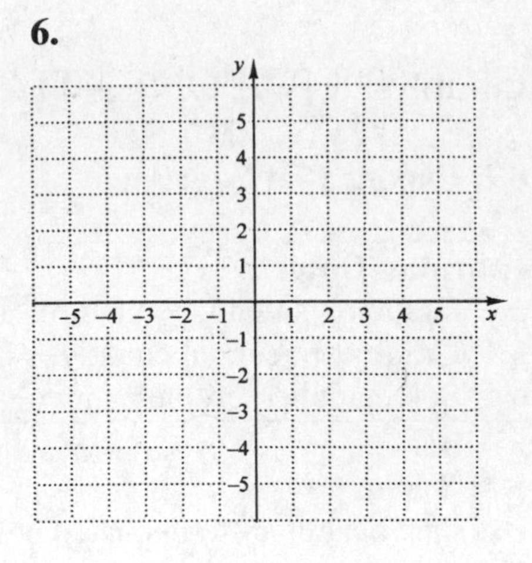

7. $f(x) = \left(\frac{1}{8}\right)^x$

7.

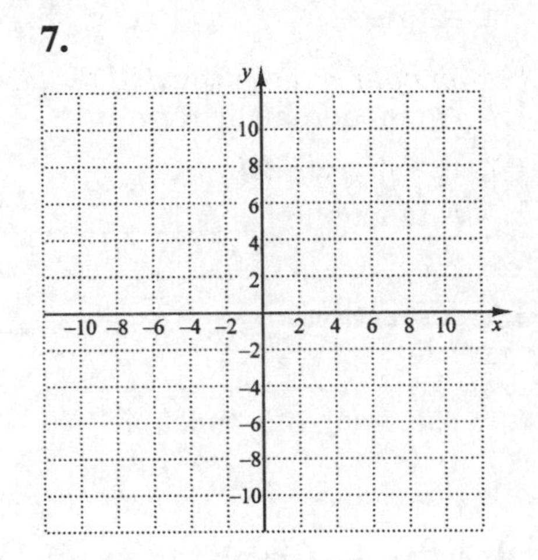

Objective b Graph exponential equations in which x and y have been interchanged.

Graph.

8. $x = 2^y$

8.

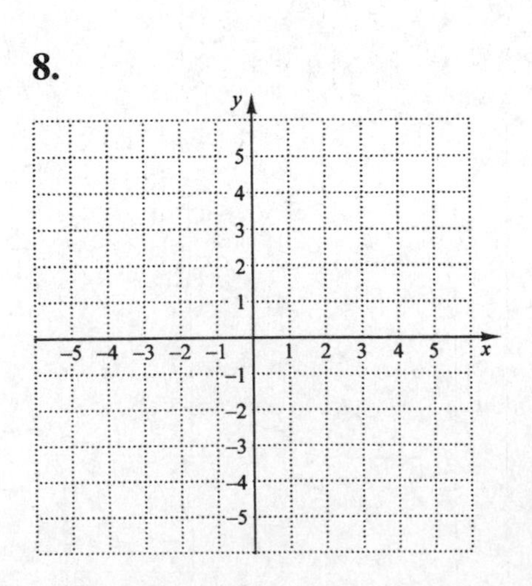

Name: Date:
Instructor: Section:

9. $x = \left(\frac{1}{2}\right)^y$

9.

Graph both equations using the same set of axes.

10. $y = 4^x$, $x = 4^y$

10.

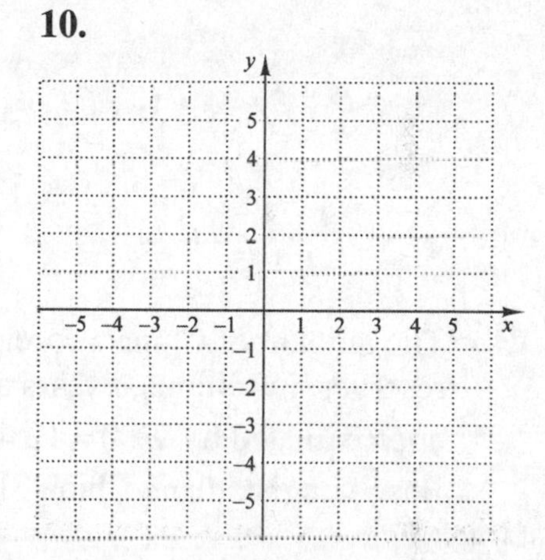

11. $y = \left(\frac{1}{3}\right)^x$, $x = \left(\frac{1}{3}\right)^y$

11.

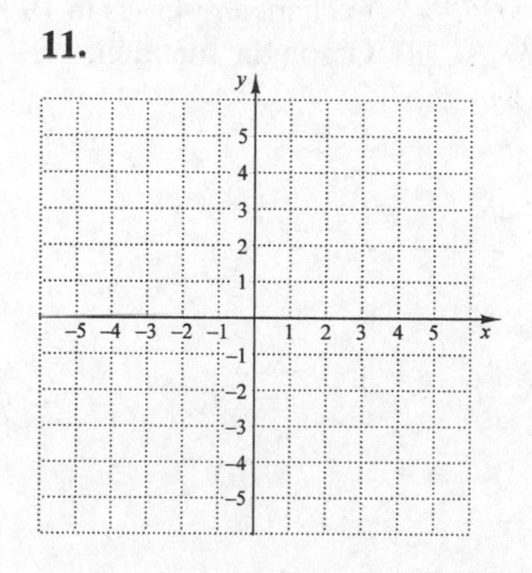

Objective c Solve applied problems involving applications of exponential functions and their graphs.

Solve.

12. The population of Ethiopia, in millions, t years after 2003, can be approximated by $P(t) = 66.56(1.221)^t$ (*Source*: Based on data from *Time Almanac 2004*).

a) Predict the population of Ethiopia in 2008 and in 2010.
b) Graph the function.

12. a)________________

b)

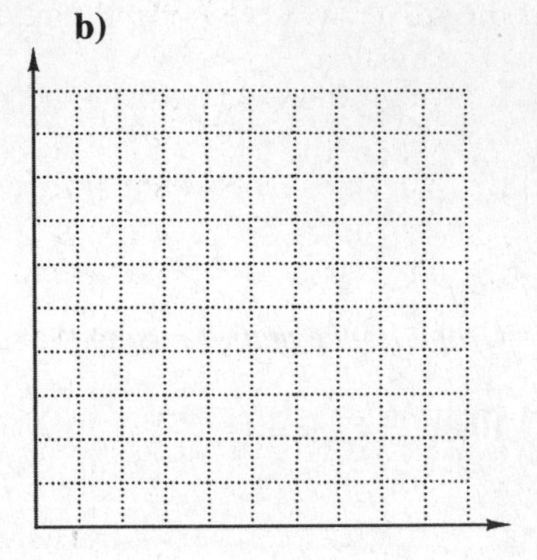

13. The amount of China's foreign-exchange reserves, in billions, t years after 1990, can be approximated by $F(t) = 12.4(1.32)^t$ (*Source*: Based on data from China Statistic Yearbook).

a) Find the amount of China's foreign-exchange reserves in 1990 and in 2005.
b) Graph the function.

13. a)________________

b)

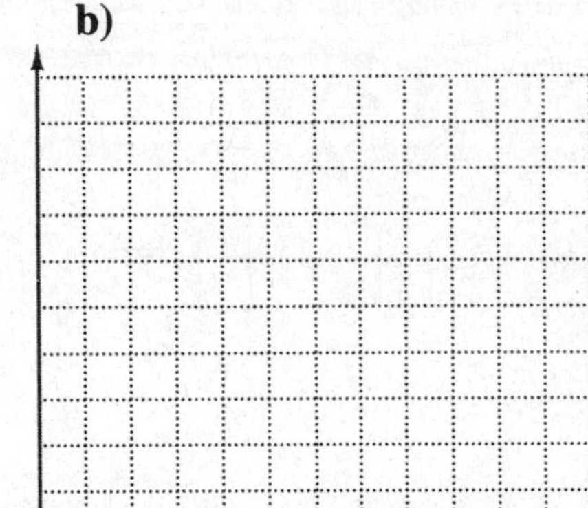

Name: Date:
Instructor: Section:

Chapter 8 EXPONENTIAL AND LOGARITHMIC FUNCTIONS

8.2 Inverse Functions and Composite Functions

Learning Objective

a Find the inverse of a relation if it is described as a set of ordered pairs or as an equation.
b Given a function, determine whether it is one-to-one and has an inverse that is a function.
c Find a formula for the inverse of a function, if it exists, and graph inverse relations and functions.
d Find the composition of functions and express certain functions as a composition of functions.
e Determine whether a function is an inverse by checking its composition with the original function.

Key Terms

Use the vocabulary terms listed below to complete each statement in Exercises 1–4.

composite **horizontal-line** **inverse** **one-to-one**

1. In a(n) __________________ function, the output depends on a variable that, in turn, depends on another variable.

2. A function *f* is __________________ if different inputs have different outputs.

3. Exchanging the first and second coordinates of all ordered pairs in a relation results in the __________________ relation.

4. A function *f* is one-to-one if it passes the __________________ test.

Objective a Find the inverse of a relation if it is described as a set of ordered pairs or as an equation.

Find the inverse of the relation.

5. $\{(2, -1), (6, 3), (6, -4), (8, 0)\}$ **5.** ______________

Find an equation of the inverse of the relation. Then complete the second table.

6. $y = 4x + 2$

x	y
-1	-2
0	2
1	6

x	y
-2	
2	
6	

6. ______________

Objective b Given a function, determine whether it is one-to-one and has an inverse that is a function.

Determine whether each function is one-to-one.

7. $f(x) = 3x - 4$ **7.** ______________

8. $f(x) = x^2 + 5$ **8.** ______________

9. $f(x) = |x + 3|$ **9.** ______________

10. $f(x) = 5^x$ **10.** ______________

Objective c Find a formula for the inverse of a function, if it exists, and graph inverse relations and functions.

Determine whether each function is one-to-one. If it is, find a formula for its inverse.

11. $f(x) = x + 1$ **11.** ______________

12. $f(x) = 5x - 2$ **12.** ______________

Name: Date:
Instructor: Section:

13. $f(x) = x^2 - 3$

13. ________________

14. $f(x) = \frac{2}{x}$

14. ________________

15. $f(x) = x^3 + 2$

15. ________________

16. $f(x) = 10$

16. ________________

Graph each function and its inverse on the same set of axes.

17. $f(x) = \frac{4}{5}x + 3$

17.

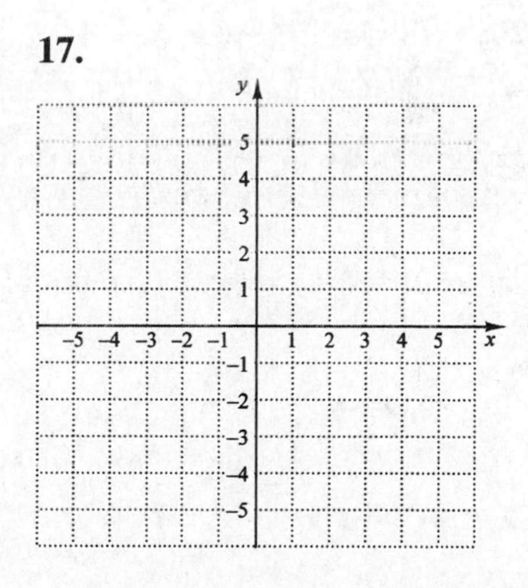

18. $f(x) = x^3 - 4$

18.

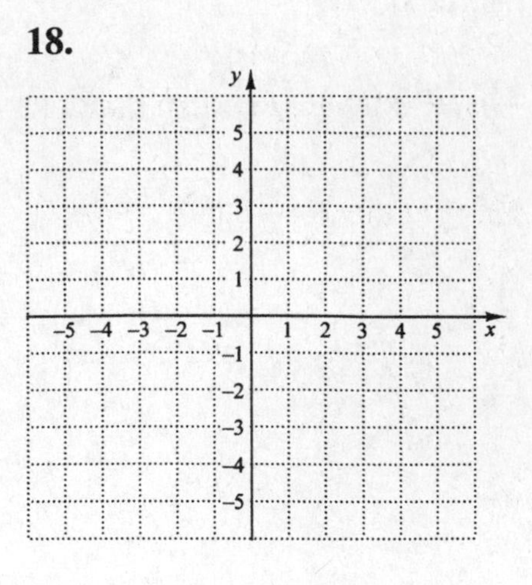

Objective d Find the composition of functions and express certain functions as a composition of functions.

Find $(f \circ g)(x)$ *and* $(g \circ f)(x)$.

19. $f(x) = 7 - 3x$

$g(x) = 0.46x + 5$

19.________________

____

20. $f(x) = 2x^2 - 6$

$g(x) = 4x + 9$

20.________________

____

21. $f(x) = 5x^2 + 2$

$g(x) = \frac{4}{x}$

21.________________

____

22. $f(x) = x^2 - 4$

$g(x) = x^2 + 4$

22.________________

____

Name: Date:
Instructor: Section:

Find $f(x)$ and $g(x)$ such that $h(x)=(f\circ g)(x)$. Answers may vary.

23. $h(x)=(10-3x)^2$

23.________________

____

24. $h(x)=\sqrt{4x+9}$

24.________________

____

25. $h(x)=\dfrac{4}{\sqrt{2x+7}}$

25.________________

____

26. $h(x)=\dfrac{1}{\sqrt{5x}}-\sqrt{5x}$

26.________________

____

Objective e Determine whether a function is an inverse by checking its composition with the original function.

27. Let $f(x)=\sqrt[3]{x+2}$. Use composition to show that $f^{-1}(x)=x^3-2$.

27.________________

____

____

____

28. Let $f(x)=\frac{1+x}{x}$. Use composition to show that $f^{-1}(x)=\frac{1}{x-1}$.

28. ______________

____

____

____

Find the inverse of the given function by thinking about the operations of the function and then reversing, or undoing, them. Then use composition to show whether the inverse is correct.

29. $f(x)=4x$

29. ______________

30. $f(x)=\frac{1}{5}x-9$

30. ______________

31. $f(x)=\frac{1}{x}$

31. ______________

32. $f(x)=\sqrt[3]{x+6}$

32. ______________

33. Size 39 shoes in France are size 7 in the United States. A function that converts shoe sizes in France to those in the United States is $f(x)=x-32$.

a) Find the shoe size in the United States that corresponds to size 37 in France.

b) Determine whether this function has an inverse that is a function. If so, find a formula for the inverse.

c) Use the inverse function to find the shoe size in France that corresponds to size 8 in the United States.

33.

a)______________

b)______________

c)______________

Name: Date:
Instructor: Section:

Chapter 8 EXPONENTIAL AND LOGARITHMIC FUNCTIONS

8.3 Logarithmic Functions

Learning Objectives
- a Graph logarithmic equations.
- b Convert from exponential equations to logarithmic equations and from logarithmic equations to exponential equations.
- c Solve logarithmic equations.
- d Find common logarithms on a calculator.

Key Terms
Use the vocabulary terms listed below to complete each statement in Exercises 1–4.

common **exponent** **log *x*** **logarithmic**

1. The inverse of an exponential function is a(n) __________________ function.

2. $\log_2 8$ is the __________________ to which we raise 2 to get 8.

3. Base-10 logarithms are called __________________ logarithms.

4. $\log_{10} x$ is abbreviated __________________.

Objective a Graph logarithmic equations.

Graph.

5. $y = \log_5 x$

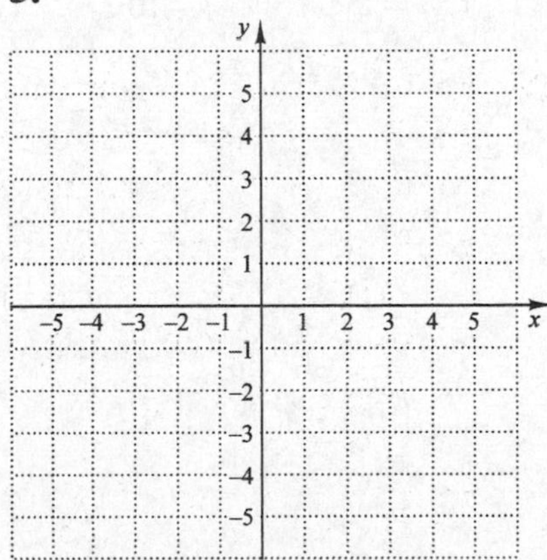

6. $f(x) = \log_{10} x$

6.

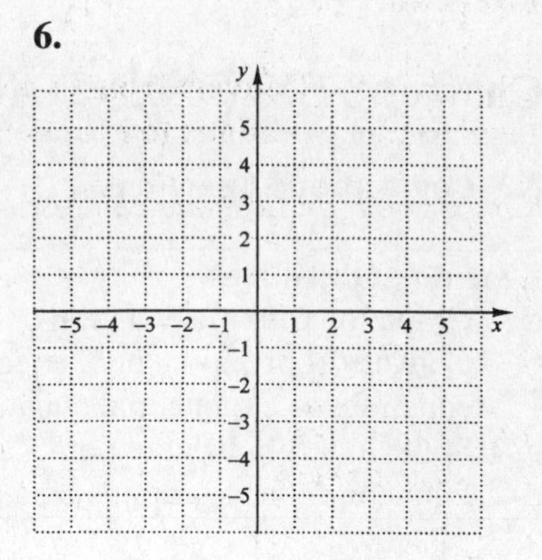

Graph both functions using the same set of axes.

7. $f(x) = 6^x, f^{-1}(x) = \log_6 x$

7.

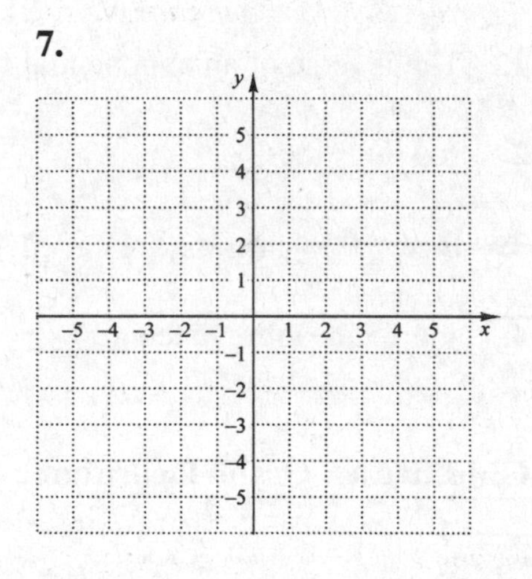

8. $f(x) = 2^x, f^{-1}(x) = \log_2 x$

8.

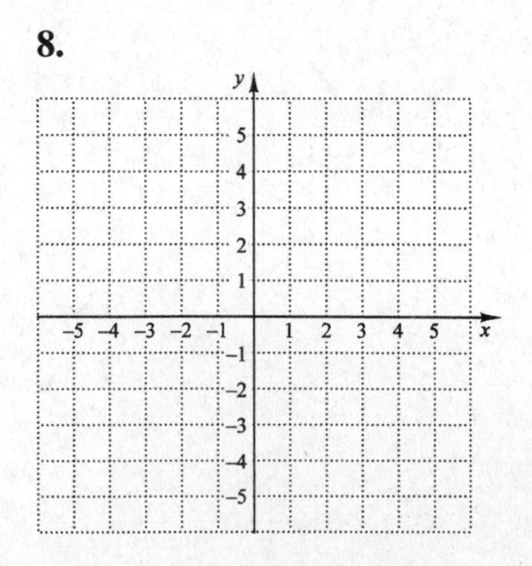

Name: Date:
Instructor: Section:

Objective b Convert from exponential equations to logarithmic equations and from logarithmic equations to exponential equations.

Convert to a logarithmic equation.

9. $27^{2/3} = 9$ **9.** ______________

10. $d^t = m$ **10.** ______________

11. $e^{-3} = 0.0498$ **11.** ______________

Convert to an exponential equation.

12. $\log_5 0.2 = -1$ **12.** ______________

13. $\log_a t = p$ **13.** ______________

14. $\log_e 0.4 = -0.9163$ **14.** ______________

Objective c Solve logarithmic equations.

Solve.

15. $\log_2 x = 4$ **15.** ______________

16. $\log_5 125 = x$ **16.** ______________

17. $\log_x 1000 = 1$ **17.** ______________

18. $\log_{81} x = \frac{1}{2}$ 18. ________

Find each of the following.

19. $\log_{10} 100{,}000$ 19. ________

20. $\log_4 64$ 20. ________

21. $\log_8 \frac{1}{8}$ 21. ________

22. $\log_9 9$ 22. ________

23. $\log_5 1$ 23. ________

Objective d Find common logarithms on a calculator.

Find the common logarithm, to four decimal places, on a calculator.

24. $\log 87{,}742.5$ 24. ________

25. $\log 0.29$ 25. ________

26. $\log(-41)$ 26. ________

27. $\log \frac{15}{36.2}$ 27. ________

Name: Date:
Instructor: Section:

Chapter 8 EXPONENTIAL AND LOGARITHMIC FUNCTIONS

8.4 Properties of Logarithmic Functions

Learning Objectives

a Express the logarithm of a product as a sum of logarithms, and conversely.
b Express the logarithm of a power as a product.
c Express the logarithm of a quotient as a difference of logarithms, and conversely.
d Convert from logarithms of products, quotients, and powers to expressions in terms of individual logarithms, and conversely.
e Simplify expressions of the type $\log_a a^k$.

Key Terms

In Exercises 1–6, match the expression with an equivalent expression from the column on the right.

1. _____ $\log_a(MN)$

2. _____ $\log_a M^p$

3. _____ $\log_a \frac{M}{N}$

4. _____ $\log_a a^k$

5. _____ $\log_a a$

6. _____ $\log_a 1$

a) k

b) 0

c) $\log_a M + \log_a N$

d) $p \cdot \log_a M$

e) $\log_a M - \log_a N$

f) 1

Objective a Express the logarithm of a product as a sum of logarithms, and conversely.

Express as a sum of logarithms.

7. $\log_2(4 \cdot 128)$ **7.** ______________

8. $\log_b(5xy)$ **8.** ______________

Express as a single logarithm.

9. $\log_a 10 + \log_a 9$ **9.** ______________

10. $\log_y A + \log_y B$ **10.** ______________

Objective b Express the logarithm of a power as a product.

Express as a product

11. $\log_a t^6$ **11.** ______________

12. $\log_c H^{-5}$ **12.** ______________

Objective c Express the logarithm of a quotient as a difference of logarithms, and conversely.

Express as a difference of logarithms.

13. $\log_3 \frac{10}{13}$ **13.** ______________

14. $\log_t \frac{u}{v}$ **14.** ______________

Express as a single logarithm.

15. $\log_a 25 - \log_a 11$ **15.** ______________

16. $\log_b 48 - \log_b 4$ **16.** ______________

Name: Date:
Instructor: Section:

Objective d Convert from logarithms of products, quotients, and powers to expressions in terms of individual logarithms, and conversely.

Express in terms of logarithms of a single variable or a number.

17. $\log_b(wxyz)$ **17.** ______________

18. $\log_a(x^{-2}yz^3)$ **18.** ______________

19. $\log_a \frac{w^3x}{yz^2}$ **19.** ______________

20. $\log_a \sqrt[4]{\frac{x^{20}y^4}{a^5z^{11}}}$ **20.** ______________

Express as a single logarithm and, if possible, simplify.

21. $5\log_x p + 4\log_x q$ **21.** ______________

22. $\log_b bt^2 - 3\log_b \sqrt[3]{t}$ **22.** ______________

23. $\log_a 3y + 5(\log_a y - \log_a z)$ **23.** ______________

24. $\log_a(5x-5)-\log_a\left(x^2-1\right)$ 24. ______________

Given $\log_a 2 = 0.356$ and $\log_a 5 = 0.827$, find each of the following.

25. $\log_a 10$ 25. ______________

26. $\log_a \frac{5}{2}$ 26. ______________

27. $\log_a \frac{1}{2}$ 27. ______________

28. $\log_a 7$ 28. ______________

Objective e Simplify expressions of the type $\log_a a^k$.

Simplify.

29. $\log_b b^{19}$ 29. ______________

30. $\log_e e^p$ 30. ______________

Solve for x.

31. $\log_7 7^6 = x$ 31. ______________

32. $\log_a a^x = 4.7$ 32. ______________

Name: Date:
Instructor: Section:

Chapter 8 EXPONENTIAL AND LOGARITHMIC FUNCTIONS

8.5 Natural Logarithmic Functions

Learning Objective

a Find logarithms or powers, base e, using a calculator.
b Use the change-of-base formula to find logarithms with bases other than e or 10.
c Graph exponential and logarithmic functions, base e.

Key Terms

Use the vocabulary terms listed below to complete each statement in Exercises 1–4.

change-of-base **domain** **natural** **range**

1. Logarithms base e are called __________________ logarithms.

2. To find a logarithm with a base other than 10 or e, we use the __________________ formula.

3. The __________________ of $f(x)=e^x$ is $\mathbb{R}$.

4. The __________________ of $f(x)=e^x$ is $(0,\infty)$.

Objective a Find logarithms or powers, base e, using a calculator.

Find each of the following logarithms or powers, base e, using a calculator. Round answers to four decimal places.

5. $\ln 8$ 5. __________________

6. $\ln 0.1$ 6. __________________

7. $\ln\left(\frac{2500}{15}\right)$ 7. __________________

8. $\ln 0$ 8. ______________

9. $\ln e^3$ 9. ______________

10. $\ln 35$ 10. ______________

11. $\ln 0.0052$ 11. ______________

12. $e^{-3.8}$ 12. ______________

13. $e^{0.587}$ 13. ______________

Objective b Use the change-of-base formula to find logarithms with bases other than *e* or 10.

Find each of the following logarithms using the change-of-base formula.

14. $\log_4 85$ 14. ______________

15. $\log_5 100$ 15. ______________

Name: Date:
Instructor: Section:

16. $\log_{0.2} 90$

16. ____________

17. $\log_{\pi} 32$

17. ____________

Objective c Graph exponential and logarithmic functions, base *e*.

Graph.

18. $f(x) = e^x + 1$

18.

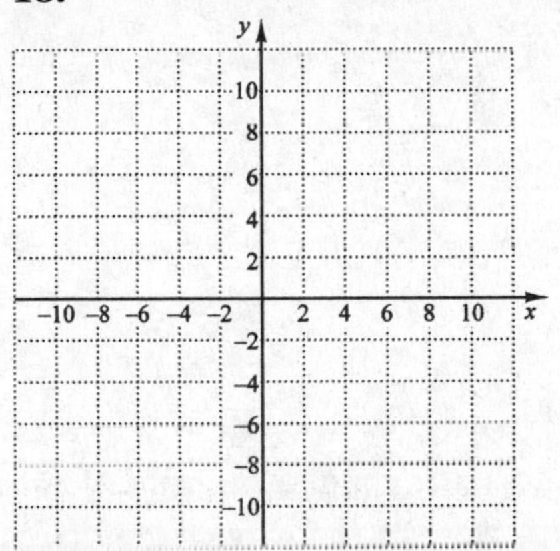

19. $f(x) = 0.4e^x$

19.

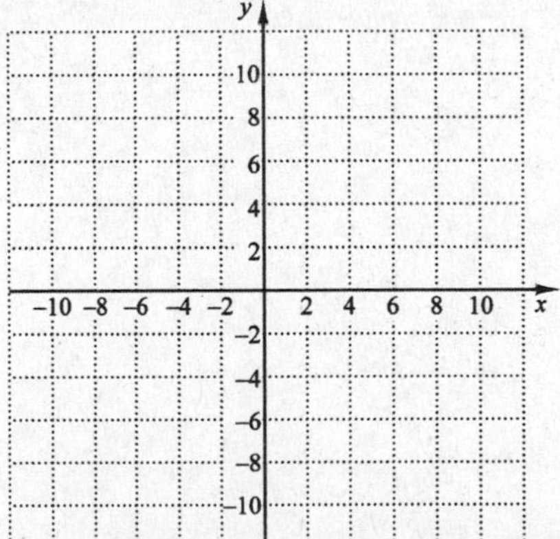

20. $g(x) = -\ln x$

20.

21. $g(x) = \ln(x - 2)$

21.

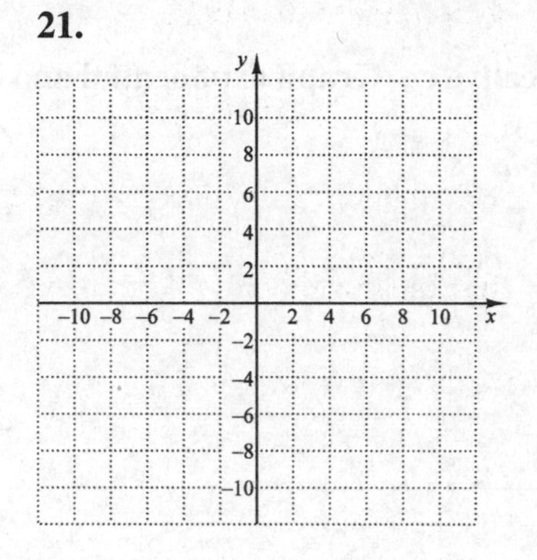

Name: Date:
Instructor: Section:

Chapter 8 EXPONENTIAL AND LOGARITHMIC FUNCTIONS

8.6 Solving Exponential and Logarithmic Equations

Learning Objective
a Solve exponential equations.
b Solve logarithmic equations.

Key Terms
Use the vocabulary terms listed below to complete each statement in Exercises 1–4. Terms may be used more than once.

exponential **logarithmic**

1. The equation $2^x = 10$ is an example of a(n) __________________ equation.

2. The equation $\log_2 x = 10$ is an example of a(n) __________________ equation.

3. The principle of __________________ equality states that for any real number *b*, where $b \neq -1, 0$, or 1, $b^x = b^y$ is equivalent to $x = y$. .

4. The principle of __________________ equality states that for any logarithmic base *a*, and for $x, y > 0$, $x = y$ is equivalent to $\log_a x = \log_a y$.

Objective a Solve exponential equations.

Solve.

5. $2^x = 11$ **5.** __________________

6. $e^t = 200$ **6.** __________________

7. $8 = 2^{x-1}$ **7.** __________________

8. $3.6^x = 25$ **8.** ________________

9. $4e^{3x} = 10$ **9.** ________________

10. $5^{3x} = 125$ **10.** ________________

11. $4^{2x-1} = 64$ **11.** ________________

12. $3^x = 10$ **12.** ________________

13. $e^{-0.05t} = 0.09$ **13.** ________________

14. $4^{x+3} = 3^{x-1}$ **14.** ________________

Name: Date:
Instructor: Section:

Objective b Solve logarithmic equations.

Solve.

15. $\log_3 x = -2$ **15.** ________________

16. $\log_9 x = \frac{1}{2}$ **16.** ________________

17. $\ln x = 9$ **17.** ________________

18. $\ln(2x+3) = 6$ **18.** ________________

19. $2\log x = 3$ **19.** ________________

20. $\log_2(9-x) = 3$ **20.** ________________

21. $\log(x+3)+\log x=1$ 21. ______________

22. $\log x-\log(x+1)=1$ 22. ______________

23. $\log_5(x+2)=3+\log_5(x-4)$ 23. ______________

24. $\log_2(x+6)+\log_2 x=4$ 24. ______________

Name: Date:
Instructor: Section:

Chapter 8 EXPONENTIAL AND LOGARITHMIC FUNCTIONS

8.7 Mathematical Modeling with Exponential and Logarithmic Functions

Learning Objectives
a Solve applied problems involving logarithmic functions.
b Solve applied problems involving exponential functions.

Key Terms
Use the vocabulary terms listed below to complete each statement in Exercises 1–6.

decay	**decibel**	**doubling time**
half-life	**growth**	**pH**

1. The ____________________ scale is used to measure the volume of a sound.

2. The ____________________ of a liquid is a measure of its acidity.

3. An exponential ____________________ model is a function of the form $P(t) = P_0 e^{kt}, k > 0$.

4. The ____________________ is the amount of time necessary for a population to double in size.

5. An exponential ____________________ model is a function of the form $P(t) = P_0 e^{-kt}, k > 0$.

6. The ____________________ is the amount of time necessary for half of a quantity to decay.

Objective a Solve applied problems involving logarithmic functions.

Solve.

7. The intensity of the sound from a large orchestra is about 6.3×10^{-3} W/m^2. How loud in decibels is this sound level? Use

$$L = 10 \cdot \log \frac{I}{I_0}, \text{ where } I_0 = 10^{-12} \text{ W/m}^2.$$

7. ______________

8. The sound level in a subway is 88 dB. What is the intensity of this sound? Use
$L = 10 \cdot \log \frac{I}{I_0}$, where $I_0 = 10^{-12}$ W/m^2.

8. ______________

9. The hydrogen ion concentration of ammonia is 3.16×10^{-11} moles per liter. Find the pH. Use
$\text{pH} = -\log\left[\text{H}^+\right]$.

9. ______________

10. The pH of lemon juice is 2.0. What is the hydrogen ion concentration of lemon juice? Use
$\text{pH} = -\log\left[\text{H}^+\right]$.

10. ______________

11. Students in a literature class took a final exam. They took equivalent forms of the exam at monthly intervals thereafter. The average score $S(t)$, in percent, after t months was found to be given by $S(t) = 72 - 15\log(t+1), t \geq 0$.

a) What was the average test score when they initially took the test, $t = 0$?
b) What was the average score after 3 months?
c) What was the average score after 12 months?

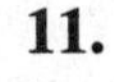

a) ______________

b) ______________

c) ______________

Name: Date:
Instructor: Section:

Objective b Solve applied problems involving exponential functions.

Solve.

12. The population of the Marshall Islands, in thousands, t years after 2003 can be approximated by $P(t) = 56.4(1.029)^t$ (*Source*: Based on data from *Time Almanac*, 2004).

a) In what year will the population reach 75,000?

b) Find the doubling time.

12.
a)____________
b)____________

13. A college loan of $45,000 is made at 4% interest, compounded annually. After t years, the amount due, A, is given by the function $A(t) = 45{,}000(1.04)^t$.

a) After what amount of time will the amount due reach $50,000?

b) Find the doubling time.

13.
a)____________
b)____________

14. Suppose that P_0 is invested in a savings account where interest is compounded continuously at 3.5% per year.

a) Using the model $P(t) = P_0e^{kt}$, express $P(t)$ in terms of P_0 and 0.035.

b) Suppose that $2000 is invested. What is the balance after 1 yr? After 2 yr?

c) When will an investment of $2000 double itself?

14.
a)____________
b)____________
c)____________

15. In 2006, the population of Uganda was 28 million and the exponential growth rate was 3% per year (*Source*: *Time Almanac*).

a) Find the exponential growth function.

b) Predict the population of Uganda in 2010.

c) When will the population of Uganda reach 40 million?

15.

a)____________

b)____________

c)____________

16. The exponential growth rate of the population of Mexico is 1.16% (*Source*: *The World Factbook*, www.cia.gov). What is the doubling time?

16. ____________

17. How old is an archaeological discovery that has lost 15% of its carbon-14? Use $P(t) = P_0e^{-0.00012t}$.

17. ____________

18. The decay rate of manganese-52 is 12.5% per day. What is its half-life? Use $P(t) = P_0e^{-kt}$.

18. ____________

Name: Date:
Instructor: Section:

Chapter 9 CONIC SECTIONS

9.1 Parabolas and Circles

Learning Objectives

a Graph parabolas.

b Use the distance formula to find the distance between two points whose coordinates are known.

c Use the midpoint formula to find the midpoint of a segment when the coordinates of its endpoints are known.

d Given an equation of a circle, find its center and radius and graph it. Given the center and the radius of a circle, write an equation of the circle and graph the circle.

Key Terms

Use the vocabulary terms listed below to complete each statement in Exercises 1–4.

center **parabola** **circle** **radius**

1. The graph of $y = ax^2 + bx + c$ is a(n) ____________________ .

2. A(n) ____________________ is the set of points in a plane that are a fixed distance from a fixed point.

3. The ____________________ of the circle $(x-h)^2 + (y-k)^2 = r^2$ is (h, k).

4. The ____________________ of the circle $(x-h)^2 + (y-k)^2 = r^2$ is r.

Objective a Graph parabolas.

Graph each equation.

5. $y = -2x^2$

5.

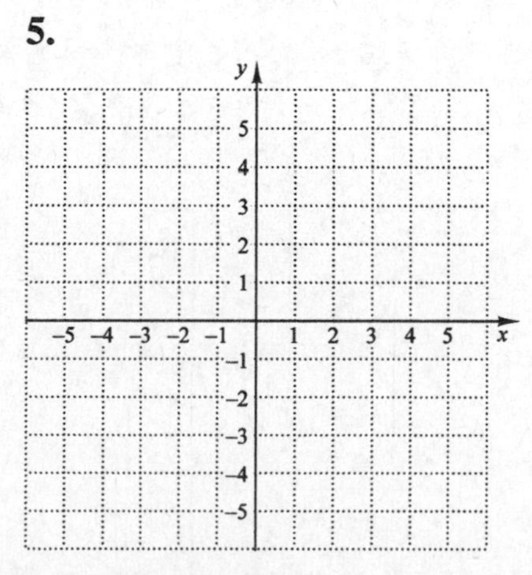

6. $x = 2 - y - y^2$

6.

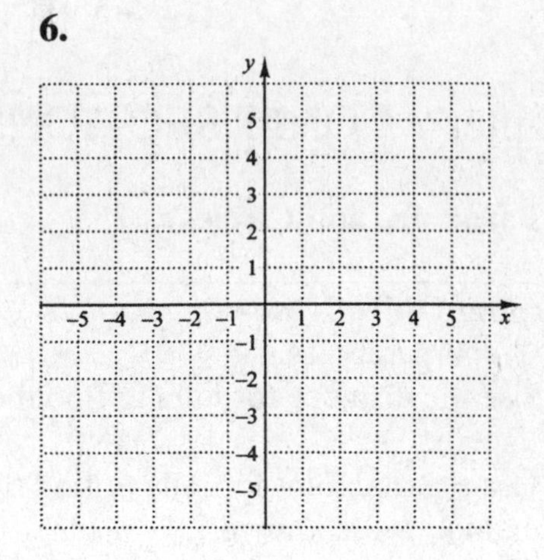

7. $x = 2y^2 - 6y + 3$

7.

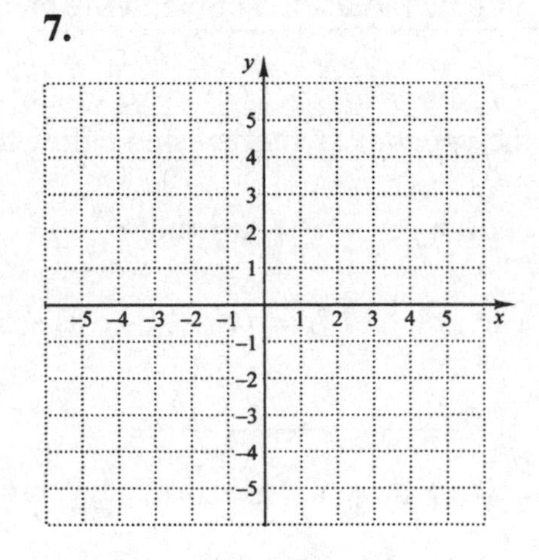

8. $y = x^2 + 4x + 4$

8.

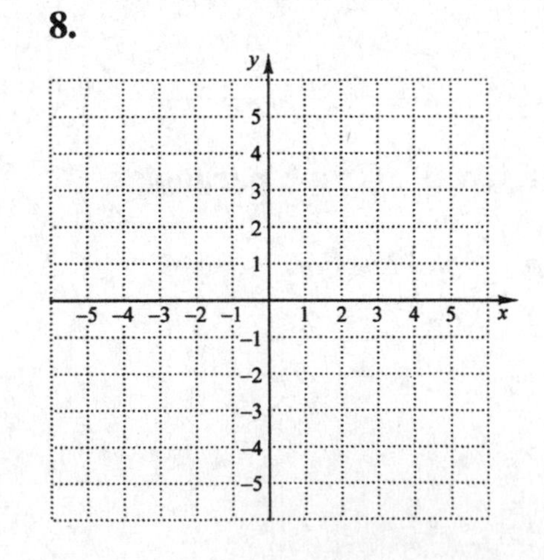

Name: Date:
Instructor: Section:

Objective b Use the distance formula to find the distance between two points whose coordinates are known.

Find the distance between each pair of points. Where appropriate, find an approximation to three decimal places.

9. $(3,10)$ and $(9,2)$ **9.**________________

10. $(2.5,-8.2)$ and $(-1.7,-8.2)$ **10.**________________

11. $\left(\frac{3}{8},\frac{1}{4}\right)$ and $\left(\frac{7}{8},\frac{5}{4}\right)$ **11.**________________

12. $(\sqrt{7},-\sqrt{11})$ and $(0,0)$ **12.**________________

13. $(-9,-15)$ and $(-11,-8)$ **13.**________________

Objective c Use the midpoint formula to find the midpoint of a segment when the coordinates of its endpoints are known.

Find the midpoint of each segment with the given endpoints.

14. $(-3,7)$ and $(1,5)$ **14.**________________

15. $(-2,-6)$ and $(12,-8)$

15.______________

16. $\left(\frac{2}{3},-\frac{5}{12}\right)$ and $\left(-\frac{1}{6},\frac{1}{3}\right)$

16.______________

17. $(\sqrt{3},-8)$ and $(\sqrt{7},13)$

17.______________

Objective d Given an equation of a circle, find its center and radius and graph it. Given the center and the radius of a circle, write an equation of the circle and graph the circle.

Find the center and the radius of each circle. Then graph the circle.

18. $x^2+y^2=16$

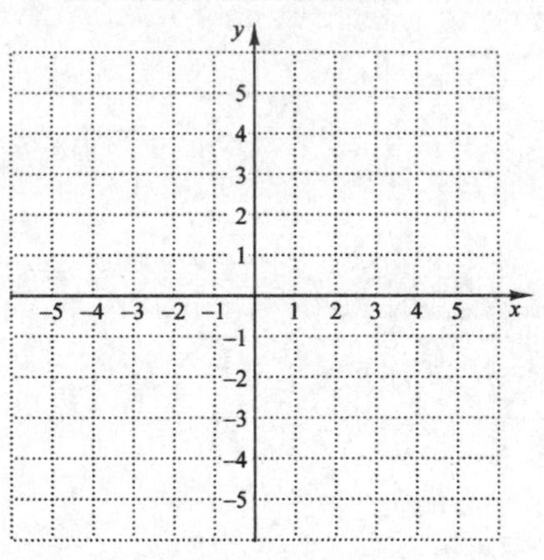

19. $(x+2)^2+(y-1)^2=4$

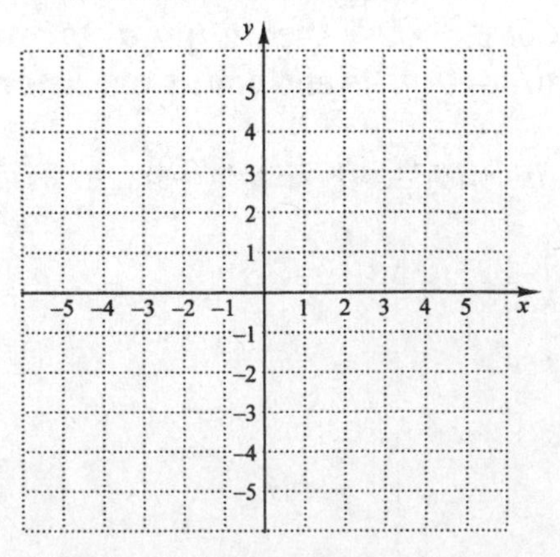

Name: Date:
Instructor: Section:

20. $(x-5)^2+(y+3)^2=12$

20.________________

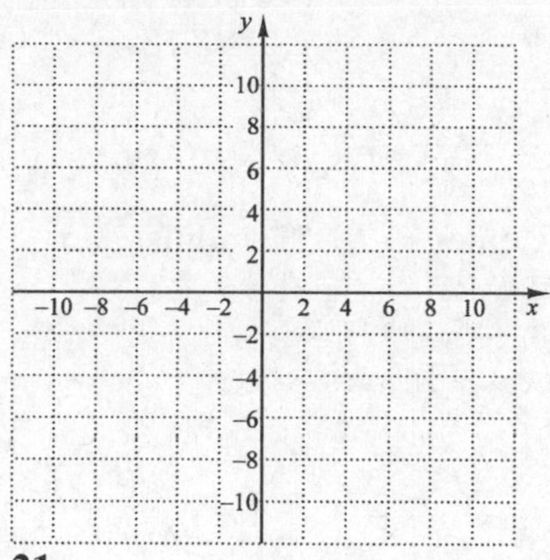

21. $x^2+y^2=5$

21.
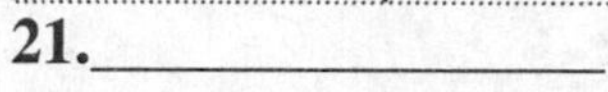

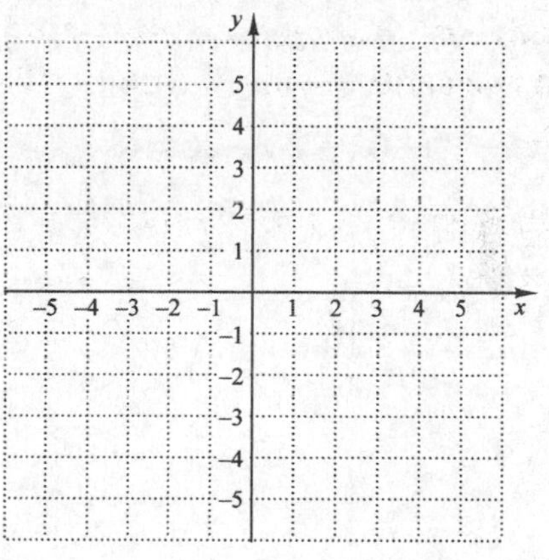

Find an equation of the circle satisfying the given conditions.

22. Center $(0,0)$, radius 9

22.________________

23. Center $(6,2)$, radius $\sqrt{6}$

23.________________

24. Center $(-10,1)$, radius $2\sqrt{2}$

24.________________

25. Center $(-2,-5)$, radius $7\sqrt{5}$

25.________________

Find the center and the radius of each circle. Then graph the circle.

26. $x^2 + y^2 + 6x - 2y - 26 = 0$

16.________________

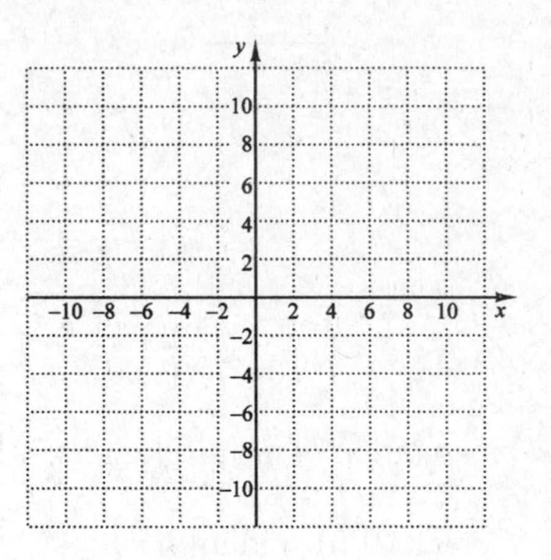

27. $x^2 + y^2 + 5y - 3 = 0$

27.________________

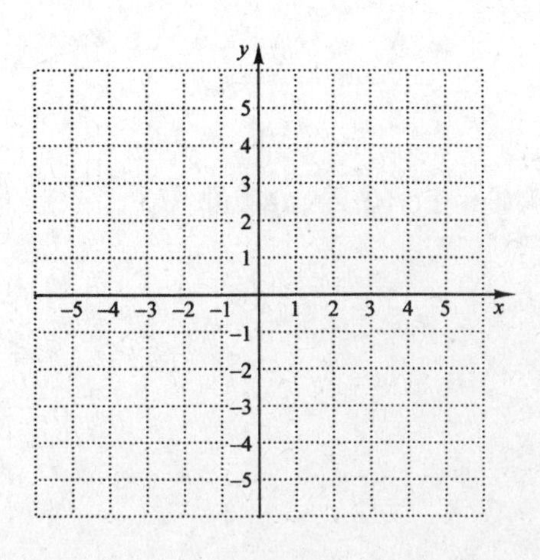

Name: Date:
Instructor: Section:

Chapter 9 CONIC SECTIONS

9.2 Ellipses

Learning Objective
a Graph the standard form of the equation of an ellipse.

Key Terms
Use the vocabulary terms listed below to complete each statement in Exercises 1–4. For each statement, consider the equation $\frac{x^2}{a^2}+\frac{y^2}{b^2}=1$.

center **ellipse** **horizontal** **vertical**

1. The graph of the equation is a(n) ____________________.

2. The ____________________ of the graph is (0, 0).

3. If $a^2 > b^2$, the ellipse is ____________________.

4. If $b^2 > a^2$, the ellipse is ____________________.

Objective a Graph the standard form of the equation of an ellipse.
Graph each of the following equations.

5. $\frac{x^2}{16}+\frac{y^2}{9}=1$

5.

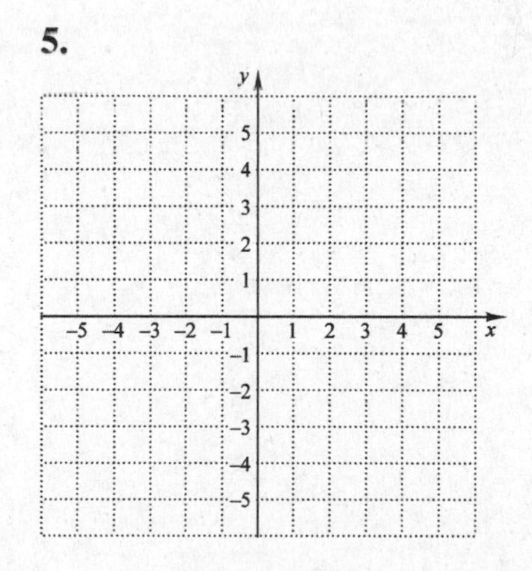

6. $\dfrac{x^2}{1}+\dfrac{y^2}{25}=1$

6.

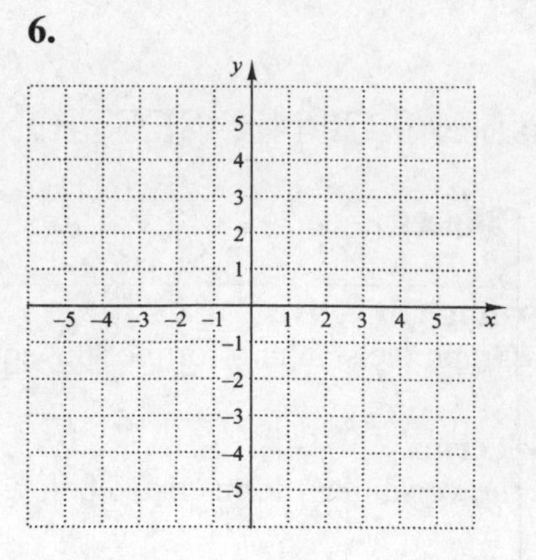

7. $25x^2+4y^2=100$

7.

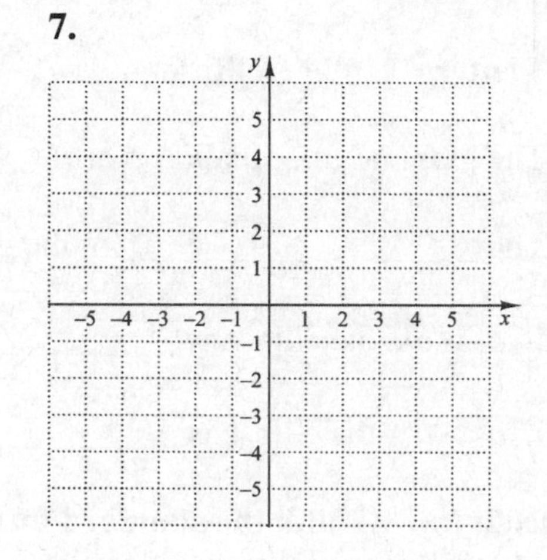

8. $3x^2+4y^2=12$

8.

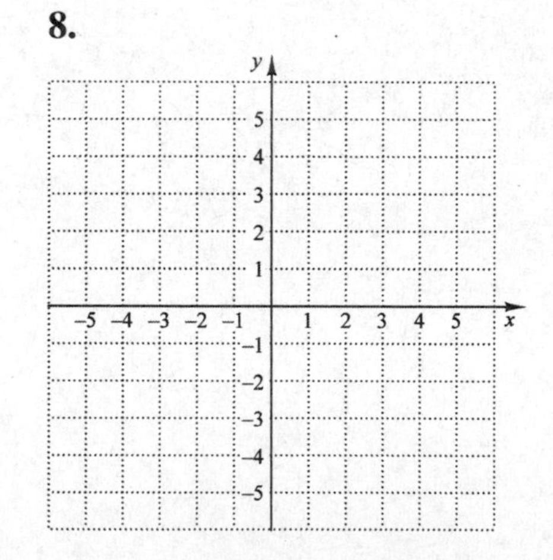

Name: Date:
Instructor: Section:

9. $5x^2+7y^2-70=0$

9.

10. $9x^2+y^2=1$

10.

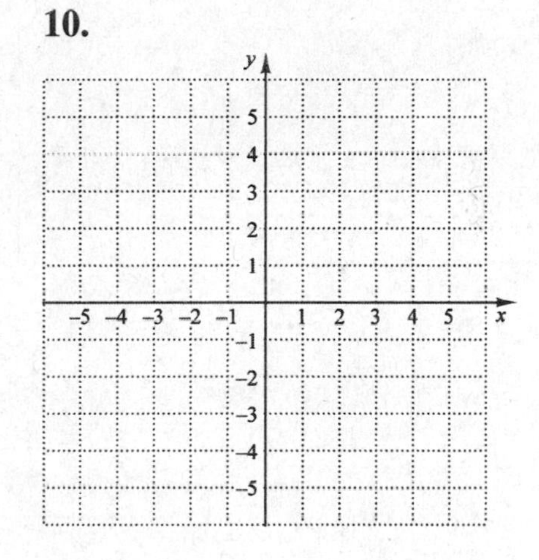

11. $\dfrac{(x-1)^2}{9}+\dfrac{(y+2)^2}{4}=1$

11.

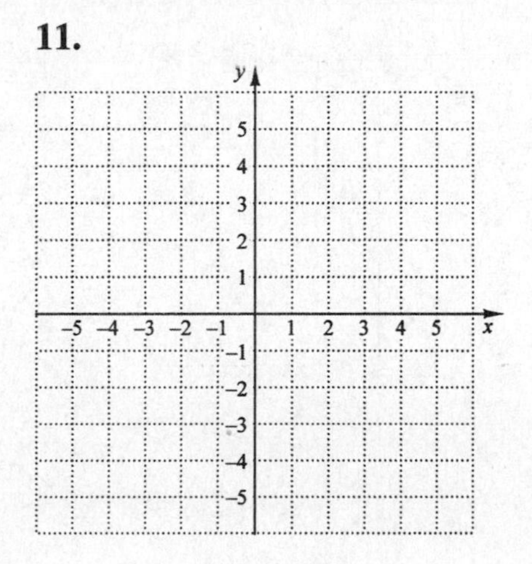

12. $\dfrac{(x+2)^2}{25}+\dfrac{(y-3)^2}{49}=1$

12.

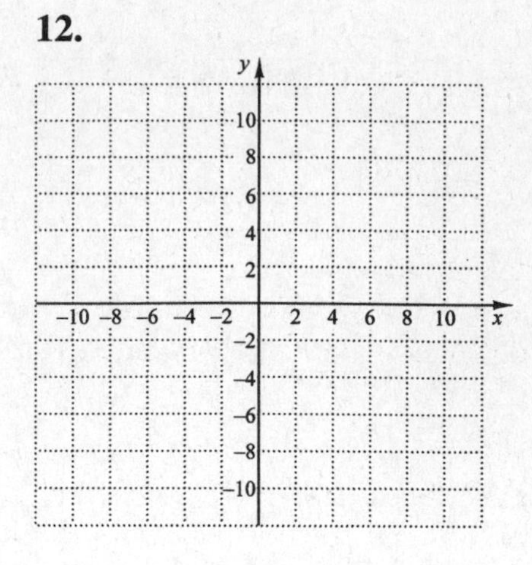

13. $15(x+3)^2+4(y-1)^2=60$

13.

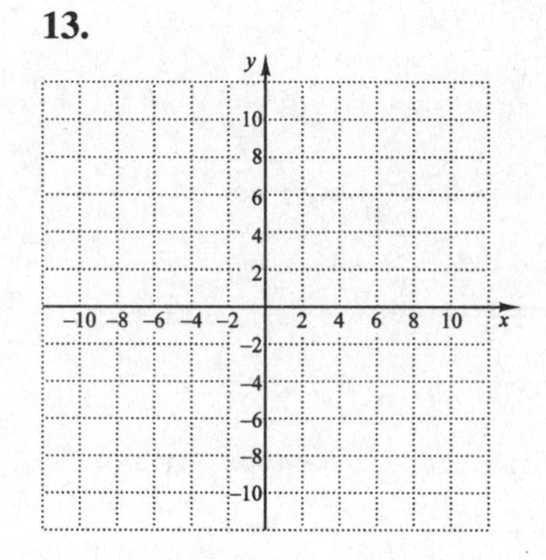

14. $2(x-2)^2+2(y+1)^2-8=10$

14.

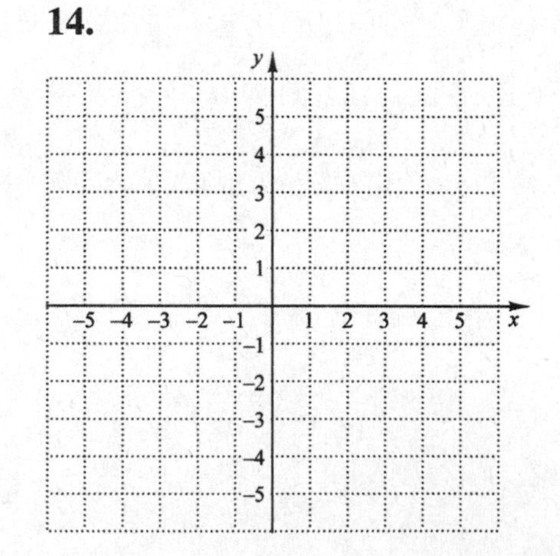

Name: Date:
Instructor: Section:

Chapter 9 CONIC SECTIONS

9.3 Hyperbolas

Learning Objectives
a Graph the standard form of the equation of an hyperbola.
b Graph equations (nonstandard form) of hyperbolas.

Key Terms
Use the vocabulary terms listed below to complete each statement in Exercises 1–5.

asymptotes	**axis**	**horizontal**
hyperbola	**vertical**	

1. The graph of the equation $xy = 8$ is a(n) ________________.

2. The line through the vertices of a hyperbola is called the ________________.

3. A hyperbola with equation $\frac{x^2}{a^2} - \frac{y^2}{b^2} = 1$ has a(n) ________________ axis .

4. A hyperbola with equation $\frac{y^2}{b^2} - \frac{x^2}{a^2} = 1$ has a(n) ________________ axis .

5. The graph of a hyperbola is constrained by ________________.

Objective a Graph the standard form of the equation of an hyperbola.

Graph each hyperbola. Label all vertices and sketch all asymptotes.

6. $\frac{x^2}{9} - \frac{y^2}{4} = 1$

6.

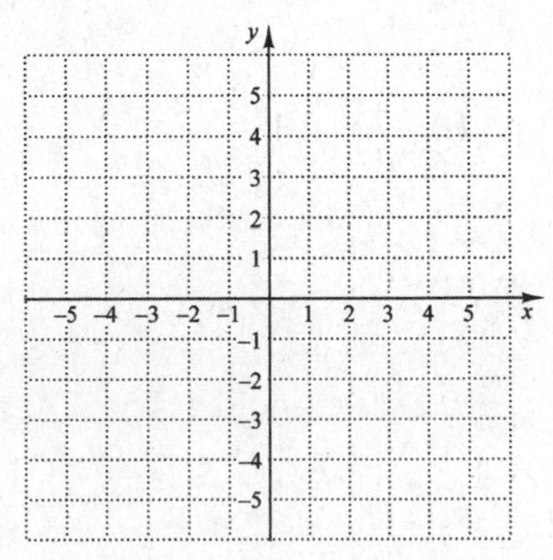

7. $x^2 - y^2 = 25$

7.

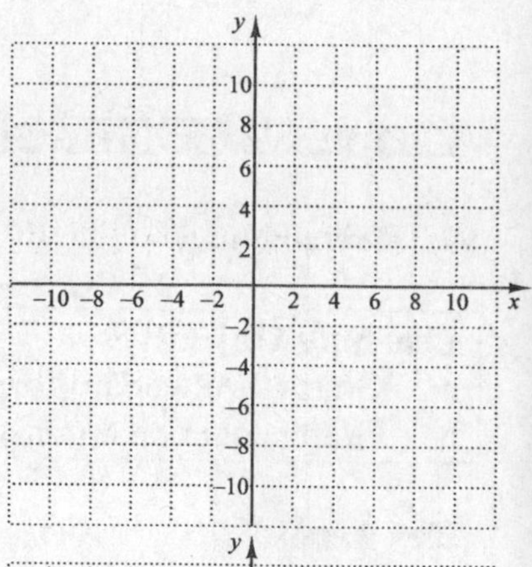

8. $25y^2 - 4x^2 = 100$

8.

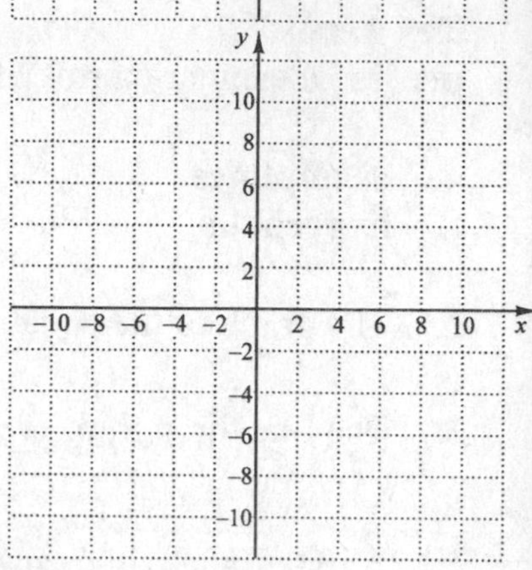

Objective b Graph equations (nonstandard form) of hyperbolas.

Graph.

9. $xy = -4$

9.

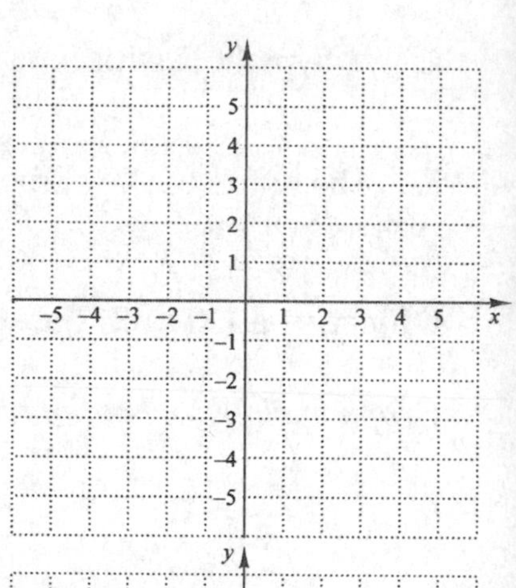

10. $xy = 3$

10.

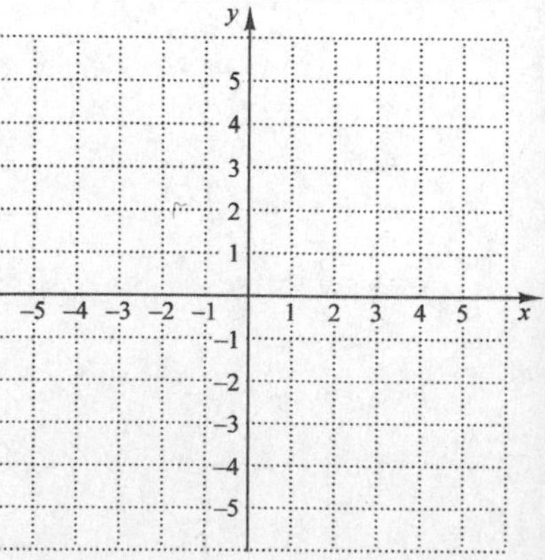

Name: Date:
Instructor: Section:

Chapter 9 CONIC SECTIONS

9.4 Nonlinear Systems of Equations

Learning Objectives
a Solve systems of equations in which at least one equation is nonlinear.
b Solve applied problems involving nonlinear systems.

Key Terms
Use the vocabulary terms listed below to complete each statement in Exercises 1–2.

elimination **substitution**

1. The system
$x - y = 5,$
$y = x^2 + 1$
is best solved using ________________.

2. The system
$x^2 - y^2 = 10,$
$2x^2 + y^2 = 3$
is best solved using ________________.

Objective a Solve systems of equations in which at least one equation is nonlinear.

Solve. Remember that graphs can be used to confirm all real solutions.

3. $x^2 + y^2 = 25,$
$x - y = 1$

3.________________

4. $a^2 = b + 10,$
$6a + b = 6$

4.________________

5. $x^2 + 4y^2 = 5,$
$x + 2y = -1$

5.______________

6. $2x + y = 2,$
$4y^2 - 2x = 1$

6.______________

7. $a + b = 5,$
$ab = 4$

7.______________

8. $x^2 + y^2 = 34,$
$y - x = 2$

8.______________

9. $x^2 - y^2 = 81,$
$x^2 + y^2 = 81$

9.______________

10. $a^2 + b^2 = 41,$
$ab = 20$

10.________________

11. $x^2 + y^2 = 4,$
$x^2 + 5y^2 = 16$

11.________________

12. $xy + y^2 = -2,$
$xy + 3y^2 = 6$

12.________________

Objective b Solve applied problems involving nonlinear systems.

Solve.

13. A rectangle has an area of 5 ft^2 and a perimeter of 12 ft. Find its dimensions.

13.________________

14. A rectangle has a perimeter of 10 cm and a diagonal of length $\sqrt{13}$ cm. Find its dimensions.

14.________________

15. A certain amount of money saved for 1 yr at a certain interest rate yielded $75 in interest. If $375 more had been invested and the rate had been 1% less, the interest would have been the same. Find the principal and the rate.

15.________________

16. Kyle made two square cakes for a reception. Find the length of each cake if the sum of their areas is 900 in^2 and the difference of their areas is 252 in^2.

16.________________

Chapter R REVIEW OF BASIC ALGEBRA

Section R.1

Key Terms

1. opposite

3. roster method

Objective a

5. $0, 9, \sqrt{36}$

7. $-7, 0, 9, \sqrt{36}$

9. $\sqrt{14}, \pi$

11. {n, u, m, b, e, r}

13. {1, 3, 5, 7}

15. $\{x \mid x < 5\}$

Objective b

17. $>$

19. $<$

21. $<$

23. $-2 \le x$

25. $4.7 < w$

27. False

Objective c

29.

31. 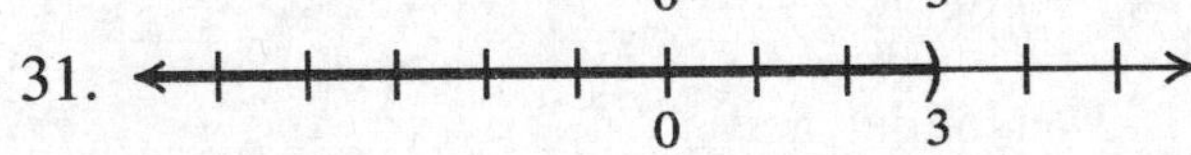

Objective d

33. 8

35. 243

37. 18.9

39. 0

Section R.2

Key Terms

1. opposites; additive

Objective a

3. -21

5. -28

7. -22

9. 1.5

11. $-\frac{1}{15}$

Objective b

13. 0
15. –5.9
17. 0

Objective c

19. –8
21. 38
23. 0
25. $-\frac{13}{10}$

Objective d

27. 36
29. 13.6
31. 1
33. 7

Objective e

35. –8
37. 60
39. 0
41. –5
43. $\frac{4}{5}$
45. $-\frac{y}{x}$
47. $-\frac{72}{35}$
49. –6000
51. –2.5
53. Not defined

Section R.3

Key Terms

1. base

Objective a

3. 7^3
5. $(4.2)^5$
7. $\frac{1}{256}$
9. $\sqrt{11}$
11. $\frac{27}{64}$

Objective b

13. 81
15. x^8
17. 5^{-4}
19. $(-15)^{-2}$

Objective c

21. 20
23. 24
25. 117
27. 34
29. 173

Section R.4

Key Terms

1. algebraic expression
3. constant
5. evaluating

Objective a

7. $x+8$, or $8+x$
9. $b \div x$, or $\frac{b}{x}$, or b/x, or $b \cdot \frac{1}{x}$
11. pq
13. $7d+10$, or $10+7d$
15. $4y+7$, or $7+4y$
17. $p-0.25p$

Objective b

19. 30
21. 4
23. 240 cm^2
25. $270

Section R.5

Key Terms

1. c
3. a
5. d

Objective a

Objective b

7. $\frac{9x}{8x}$
9. $\frac{15}{8}$

Objective c

11. $4+x$
13. $b+ax, b+xa$, or $xa+b$
15. $9\cdot(a\cdot b)$
17. $(8\cdot y)\cdot x, y\cdot(x\cdot 8), (y\cdot x)\cdot 8$; others are possible
19. $5ab-10ac-20ad$
21. $9(a+2b)$
23. $\frac{1}{4}b(3a+b)$
25. $z(x+1)$

Section R.6

Objective a

1. $30w$
3. $11s$
5. $1.26x+1.87y$
7. $-\frac{5}{6}a+\frac{5}{6}b-29$

Objective b

9. $-a-b-c$
11. $-90+39t$
13. $11x-18$
15. $-21z+24$
17. $2x+39$
19. $-13t+\frac{1}{3}w+10$
21. –1936

Section R.7

Key Terms

1. exponential notation

Objective a

3. u^{13}
5. $(5s)^{16}$
7. x^6y^3
9. $1296r^2$
11. $4m^3$

Objective b

13. 3^8
15. $\frac{1}{x^3y^6}$

17. $\frac{b^6c^{12}}{a^4}$

19. $\frac{x^{10}}{9}$

21. $\frac{w^9z^3}{x^3y^6}$

23. -3^{2a-2}

25. $\frac{x^{20}}{10,000y^{24}z^{12}}$

Objective c

27. 9.6×10^{11}
29. 7.7×10^{14}
31. 0.000000421
33. 0.000000000606
35. 6.956×10^4
37. 4.0×10^{-20}
39. 4.643×10^{16}
41. 9.53×10^{-3} lb

Chapter 1 SOLVING LINEAR EQUATIONS AND INEQUALITIES

Section 1.1

Key Terms

1. equation
3. addition principle
5. inverse
7. infinitely many solutions

Objective a

9. No
11. Yes
13. No

Objective b

15. –5
17. –22

Objective c

19. 7
21. –616
23. –8

Objective d

25. –11
27. 3
29. 6
31. $\frac{17}{5}$
33. 0
35. All real numbers

37. 2
39. 8
41. 0
43. 5

Section 1.2

Key Terms

1. formula

Objective a

3. $g = \frac{A}{q}$
5. $r^2 = \frac{A}{4\pi}$
7. $w = \frac{4k}{c}$
9. $y = \frac{e}{kr}$
11. $k = \frac{r - 3u}{3}$
13. $B = \frac{K}{1 + vy}$
15. $m = \frac{3}{8}D + 52$
17. $B = \frac{P}{1 + ry}$
19. 48 mg
21. 1769 cal

Section 1.3

Key Terms

1. familiarize
3. solve
5. state

Objective a

7. 64 women
9. 22, 23, 24
11. length: 30 ft; width: 14 ft
13. $25.50
15. $650

Objective b

17. 2 h
19. 12 min

Section 1.4

Key Terms

1. inequality

3. equivalent

Objective a

5. (a) No; (b) no; (c) yes; (d) yes; (e) no

Objective b

7. (–3, 6)

9. $(-\infty, -4)$

Objective c

11. $\{x|x \le 3\}$, or $(-\infty, 3]$

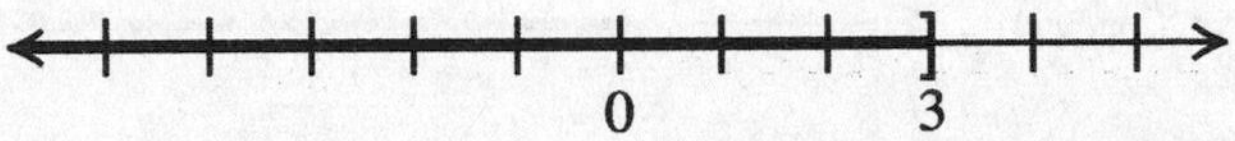

13. $\{x|x \ge 6\}$, or $[6, \infty)$

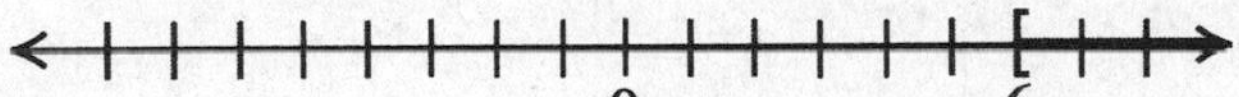

15. $\{z|z > 0\}$, or $(0, \infty)$

17. $\left\{x \middle| x > \frac{2}{21}\right\}$, or $\left(\frac{2}{21}, \infty\right)$

19. $\{x|x \le 5\}$, or $(-\infty, 5]$

21. $\{x|x \le 11\}$, or $(-\infty, 11]$

23. $\{x|x \le 6\}$, or $(-\infty, 6]$

25. $\left\{y \middle| y \le -\frac{69}{8}\right\}$, or $\left(-\infty, -\frac{69}{8}\right]$

Objective d

27. $2.99

29. $\{S|S > \$5100\}$

Section 1.5

Key Terms

1. compound inequalities

3. intersection

5. union

7. $\{5, 7, 8, 9, 10, 11, 13\}$

9. $\{f, g\}$

11. $\{2, 3, 5, 10\}$

Objective a

13. $(-2,6]$ 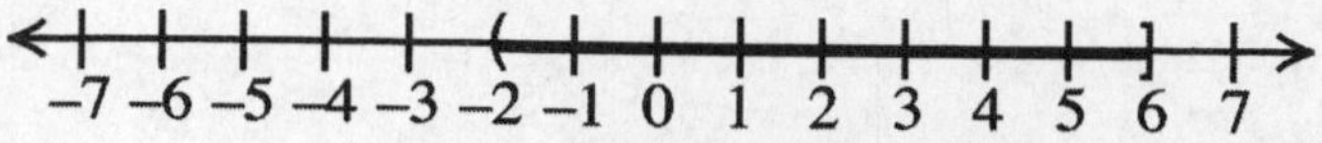

15. $\{x|1 \le x < 8\}$, or $[1,8)$ 

17. $\{x|-13 \le x \le 19\}$, or $[-13,19]$

Objective b

19. $(-\infty,0)\cup(5,\infty)$ 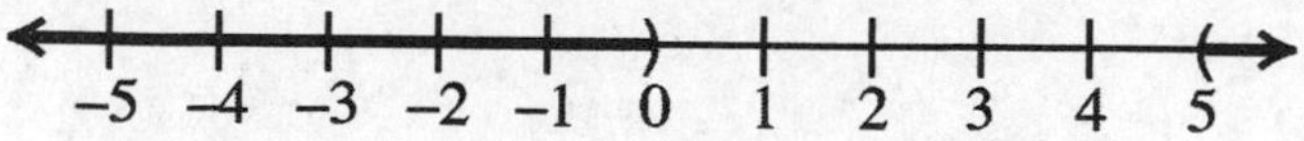

21. $[-3,\infty)$ 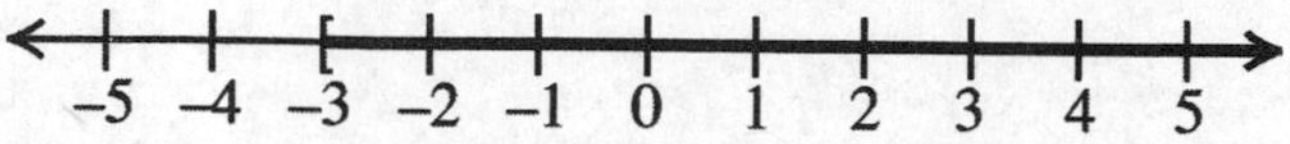

23. $\{a|a > 3\}$, or $(3,\infty)$ 

25. $\{m|m \le -3 \text{ or } m > -2\}$, or $(-\infty,-3]\cup(-2,\infty)$

Objective c

27. $\{w|144.1 \text{ lb} \le w \le 194.0 \text{ lb}\}$

Section 1.6

Key Terms

1. p
3. $-p$
5. $-p, p$

Objective a

7. $5x^2$
9. $19|z|$
11. $2|a|$
13. $9x^4$

Objective b

15. 83
17. $\frac{4}{3}$

Objective c

19. $\{0\}$

21. $\left\{-\frac{3}{4},\frac{13}{4}\right\}$

23. $\varnothing$

25. $\left\{-\frac{16}{5},\frac{16}{5}\right\}$

Objective d

27. $\left\{-7,\frac{5}{3}\right\}$

29. $\left\{-\frac{11}{3},\frac{3}{7}\right\}$

Objective e

31. $\{x|x\leq -3\ \ or\ \ x\geq 3\}$, or $(-\infty,-3]\cup[3,\infty)$

33. $\{x|-10<x<4\}$, or $(-10,4)$

35. $\left\{x\middle|x\leq -\frac{9}{2}\ \ or\ \ x\geq \frac{7}{2}\right\}$, or $\left(-\infty,-\frac{9}{2}\right]\cup\left[\frac{7}{2},\infty\right)$

37. $\left\{x\middle|x\leq -\frac{1}{4}\ \ or\ \ x\geq \frac{29}{4}\right\}$, or $\left(-\infty,-\frac{1}{4}\right]\cup\left[\frac{29}{4},\infty\right)$

Chapter 2 GRAPHS, FUNCTIONS, AND APPLICATIONS

Section 2.1

Key Terms

1. axes
3. coordinates
5. graph

Objective a

7.

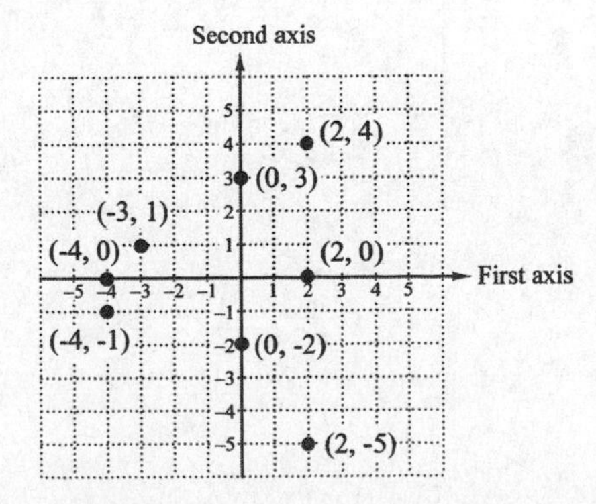

Objective b

9. Yes

11. $\dfrac{y = 2x - 3}{-1 \; ? \; 2 \cdot 1 - 3}$
    $\quad | -1$ True

    $\dfrac{y = 2x - 3}{5 \; ? \; 2 \cdot 4 - 3}$
    $\quad | \; 5$ True

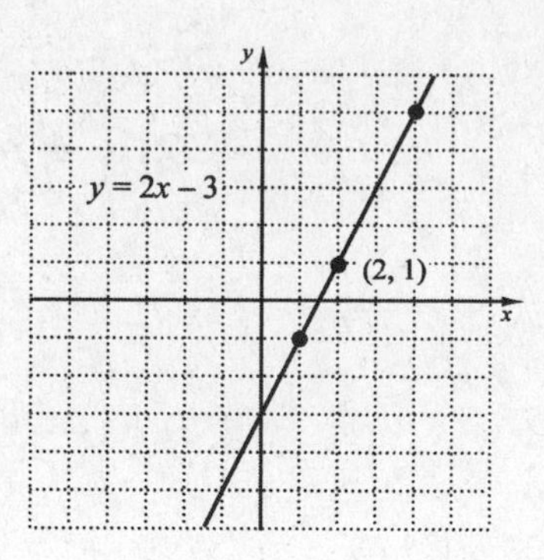

Objective c

13.

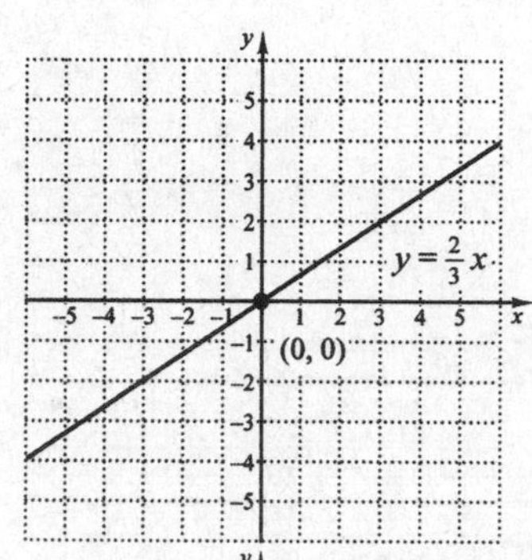

15.

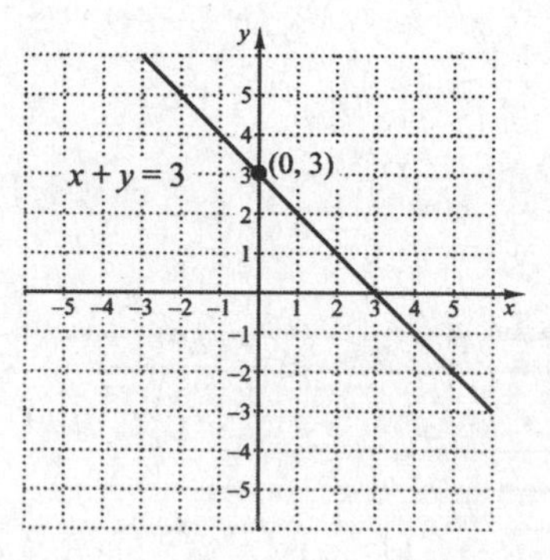

17.

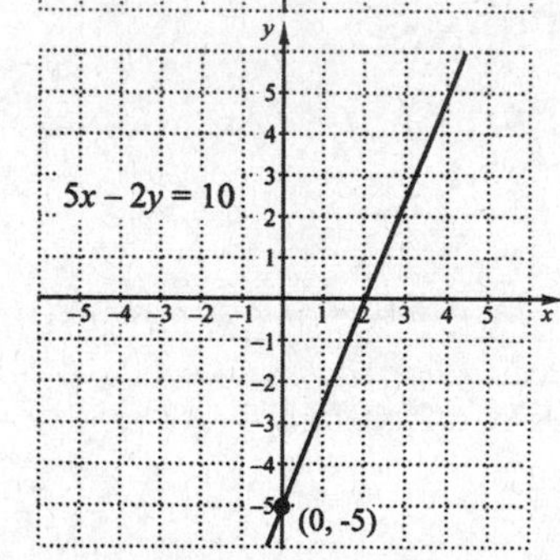

Objective d

19.

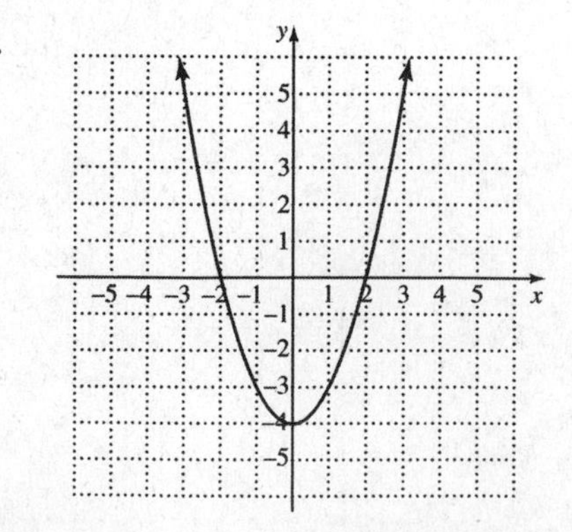

Section 2.2

Key Terms

1. input
3. outputs
5. relation

Objective a

7. No

Objective b

9. 29; −13; 46.5
11. 0; 6; 6

Objective c

13.

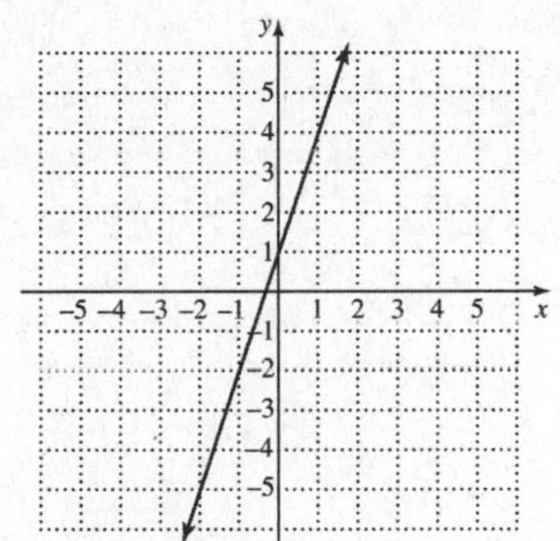

15.

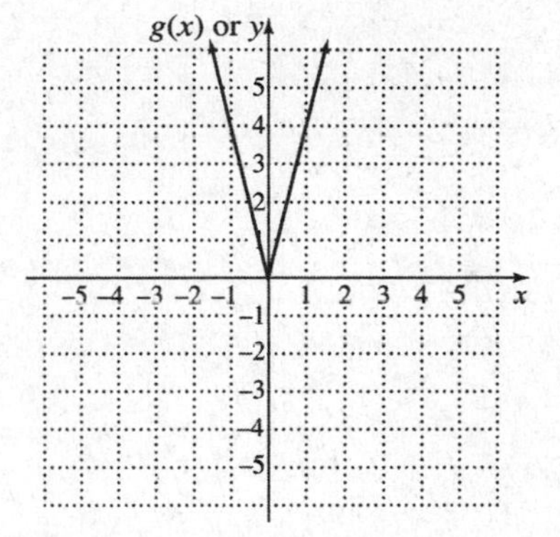

17.

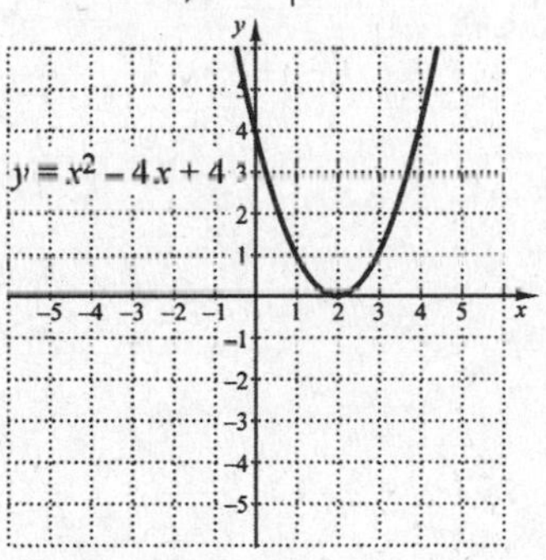

Objective d

19. Yes
21. Yes

Objective e

23. 16 words

Section 2.3

Key Terms

1. domain
3. relation

Objective a

5. a) -3; b) $\{x \mid -2 \le x \le 5\}$; c) -1; d) $\{x \mid -3 \le x \le 2\}$
7. a) -3; b) $\{-5,-4,-3,-2,-1,0,1,2,3,4,5\}$; c) -1; d) $\{-5,-4,-3,-2,-1,0,1,2,3,4,5\}$
9. $\mathbb{R}$
11. $\left\{x \,\middle|\, x \text{ is a real number } \textit{and } x \ne \frac{5}{7}\right\}$ or $\left(-\infty, \frac{5}{7}\right) \cup \left(\frac{5}{7}, \infty\right)$

Section 2.4

Key Terms

1. *m*
3. up
5. rise
7. grade

Objective a

9. $\frac{4}{5}; \left(0, -\frac{9}{5}\right)$
11. $-\frac{2}{3}; (0,3)$
13. $\frac{5}{3}; (0,3)$

Objective b

15. $\frac{2}{3}$
17. 3
19. $\frac{23}{35}$

Objective c

21. 6%
23. $-\frac{2}{3}$ gallon per year

Section 2.5

Key Terms

1. parallel
3. vertical
5. x-intercept
7. 0

Objective a

9.

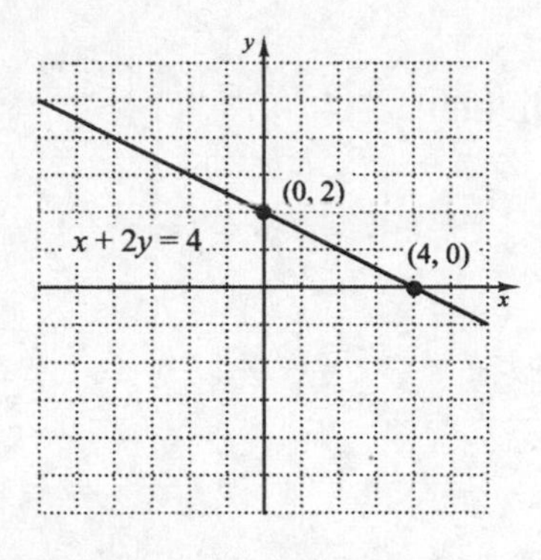

11.

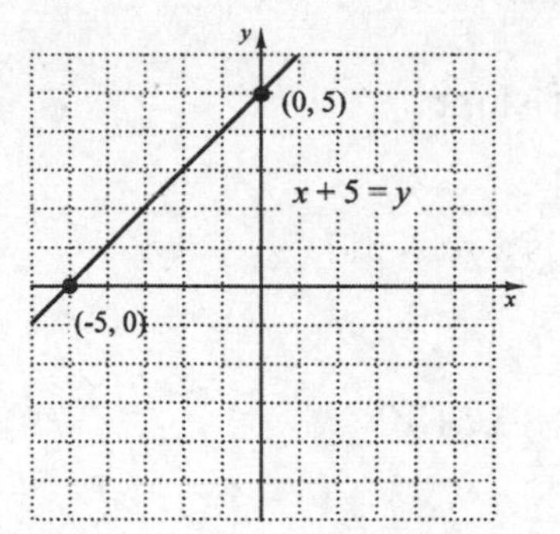

Objective b

13.

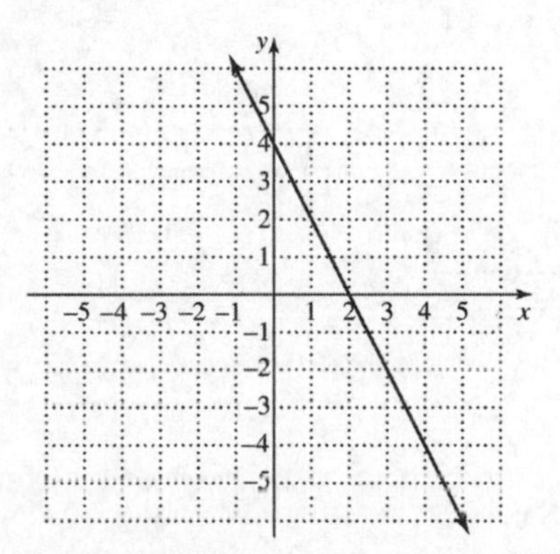

Objective c

15. Not defined

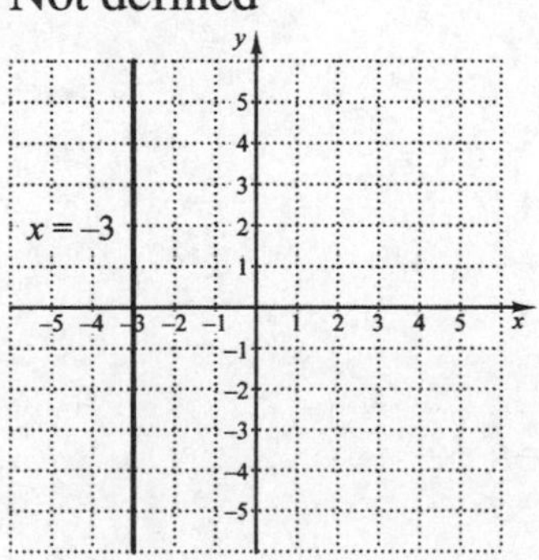

17. $m = 0$

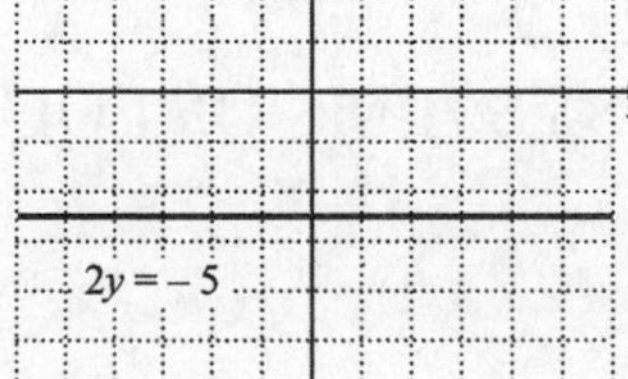

Objective d

19. Yes
21. No
23. No
25. Yes

Section 2.6

Key Terms

1. point-slope
3. perpendicular
5. slope

Objective a

7. $y=-15x+12$
9. $f(x)=-\frac{5}{6}x-8$

Objective b

11. $y=-3x-12$
13. $y=2x-11$

Objective c

15. $y=-x+11$
17. $y=\frac{4}{5}x+\frac{19}{5}$

Objective d

19. $y=4x-9$
21. $y=8x+4$
23. $y=\frac{7}{6}x+8$

Objective e

25. a) $c(t)=\frac{35}{3}t+1250$; b) 1460 students; c) 2022

Chapter 3 SYSTEMS OF EQUATIONS

Section 3.1

Key Terms

1. system
3. consistent

5. dependent

Objective a

7. $(3, 2)$; consistent; independent

9. (–2,–1); consistent; independent

11. Infinitely many solutions; consistent; dependent

13. $\left(\frac{1}{2}, -1\right)$; consistent; independent

Section 3.2

Key Terms

1. algebraic

3. solve

Objective a

5. $(2,3)$

7. $(1,1)$

9. $(-7,1)$

11. $\left(\frac{1}{2}, -\frac{3}{4}\right)$

13. $(6,11)$

15. $(0,-2)$

Objective b

17. Length: 36 ft; width: 26 ft

Section 3.3

Key Terms

1. elimination

3. false

Objective a

5. $(7,4)$

7. $(-3,-12)$

9. No solution

11. $\left(\frac{1}{3}, -2\right)$

Objective b

13. Two pointers: 32; three pointers: 9

15. $20°$, $70°$

Section 3.4

Key Terms

1. motion
3. rate

Objective a

5. Adults: 250; youth: 325
7. Soup: $2.60; tortilla wrap: $3.80
9. Tropical Punch: 12 L; Caribbean Spring: 6 L

Objective b

11. 3 hr
13. 20 mph
15. In $\frac{5}{6}$ hr, or in 50 min

Section 3.5

Objective a

1. $(3,0,1)$
3. $(5,1,-1)$
5. $\left(\frac{1}{2},-\frac{1}{2},\frac{1}{3}\right)$
7. $(2,-3,8)$

Section 3.6

Objective a

1. 36, 19, 23
3. Grapes: 60 calories; melon: 24 calories; pear: 40 calories
5. One-dip cones: 84; two-dip cones: 52; three-dip cones: 64
7. Schools: $5.6 million; roads: $1.2 million; law enforcement: $0.8 million

Section 3.7

Key Terms

1. linear inequality
3. test point

Objective a

5. Yes

Objective b

7.

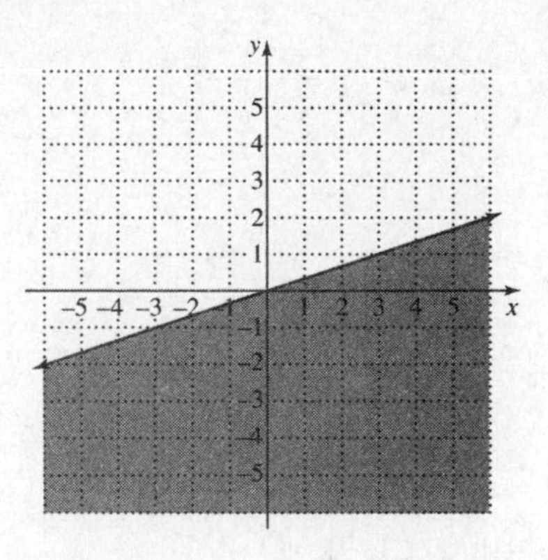

9.

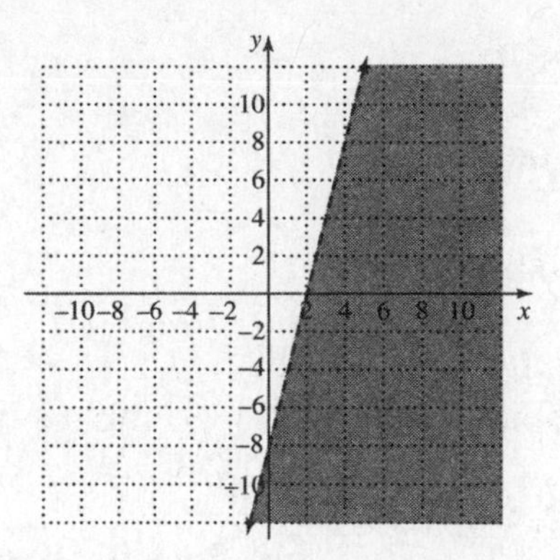

Objective c

11.

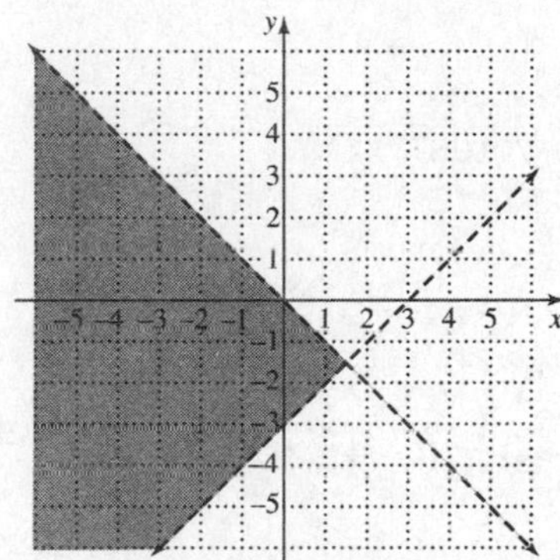

13.

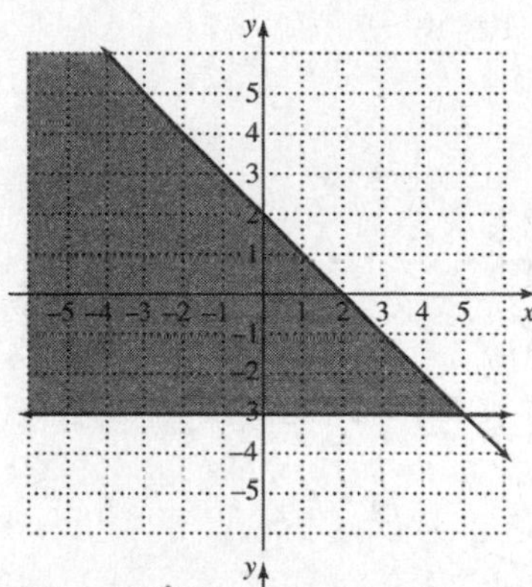

15.

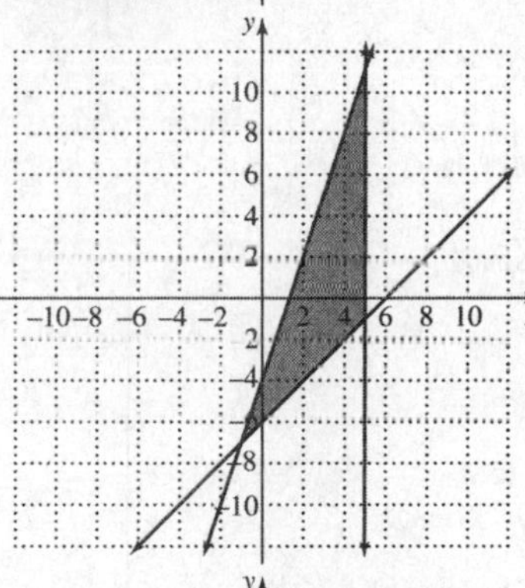

17.

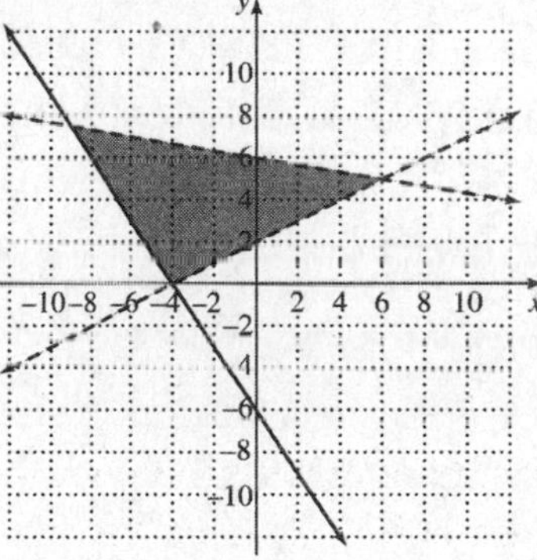

19.

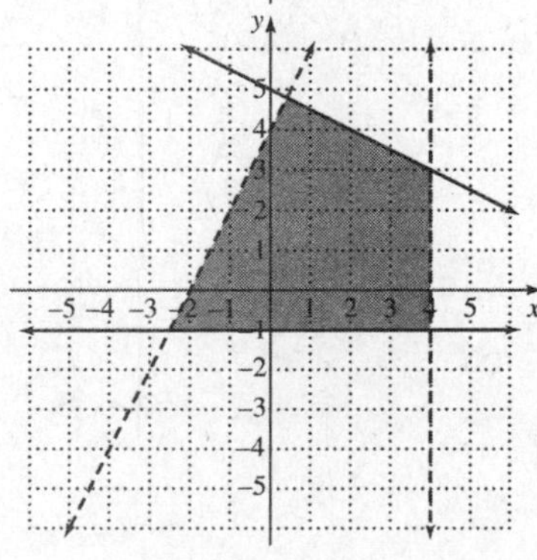

Chapter 4 POLYNOMIALS AND POLYNOMIAL FUNCTIONS

Section 4.1

Key Terms

1. polynomial
3. degree

Objective a

5. $-6b^5 - 5b^3 + 3b + 11$
7. $8 + 3x - 3x^6 + 2x^8$
9. a) $-4w^6, -7w^5, 5w^4, 5w, -9$; b) 6, 5, 4, 1, 0; c) 6; d) $-4w^6$; e) –4; f) –9

Objective b

11. -15
13. 155; 5
15. 4080
17. About 8 words

Objective c

19. $x^6 + x^5$
21. $-9x^5 + 8x$
23. $-x + 7$
25. $-1.9x^3 - 0.3x^2 - 4.5x + 62$
27. $-\frac{1}{2}x^4 + \frac{1}{4}x^3 + x^2$

Objective d

29. $-x^2 - 7x + 17$
31. $5.8x^3 + 8.4x^2 - 3.3x - 92$
33. $\frac{1}{3}x^3 - \frac{3}{4}x$

Section 4.2

Objective a

1. $12x^9$
3. $x^8 + x^3$
5. $x^2 + 6x + 8$
7. $x^3 - 6x^2 - 49$
9. $4y^4 - 9y^3 - 30y^2 + 72y - 16$
11. $x^4 - x^2 - 6x - 9$

Objective b, c

13. $x^2 + 16x + 64$
15. $a^2 - 9a + \frac{81}{4}$

17. $25x^6 - 90x^3 + 81$

19. $r^{18}t^{14} - 18r^9t^7 + 81$

Objective d

21. $p^2 - \dfrac{1}{25}$

23. $36x^{18} - 9$

25. $v^4 - t^2 j^2$

27. $x^2 - y^2 - 2y - 1$

Objective e

29. a) $t^2 - 9t + 14$ b) $5h + 2hw + h^2$

31. a) $-2p^2 - 13p - 10$ b) $-2h^2 - 4hw - 5h$

Section 4.3

Objective a

1. $y^3(y+3)$

3. $4t^2(t-5)$

5. $6x^2y^4(x-3y)$

7. $5x^2yz^4(3x^3z - 2y^3z^2 + 5xy^2)$

9. $-6(x^2 - 6x - 8)$

11. $-2(y^4 - 6s^3)$

13. $-(t^4 + t^3 - 2t - 13)$

15. $P(x) = x(10x - 9)$

Objective b

17. $2(p-2)(p+5)$

19. $(u+w)(v+t)$

21. $(x-5)(x^2-2)$

23. $a^2(a-1)(a^2+5)$

25. $(s^2+3)(5s^2+7)$

Section 4.4

Objective a

1. $(x+2)(x+9)$

3. $x(x+2)(x-6)$

5. Not factorable

7. $-(n-6)(n+7)$, or $(-n+6)(n+7)$, or $(6-n)(n+7)$

9. $(a+2b)(a-11b)$

11. $(x^2+81)(x^2+1)$

13. $(x^4-10)(x^4+3)$

15. $(1-t^{12})(9+t^{12})$, or $-1(t^{12}-1)(t^{12}+9)$

Section 4.5

Objective a, b

1. $5(n-2)(n+7)$
3. $(5x+3)(2x-1)$
5. $(10t+3)(4t-5)$
7. $m^3(7m-1)(5m+2)$
9. $-2(3z+1)(3z-5)$
11. $ab(4a+5)(2a-3)$
13. $3t^2(5t+2)(2t-5)$
15. $(5c-2d)(5c-2d)$, or $(5c-2d)^2$

Section 4.6

Key Terms

1. factored completely
3. perfect-square trinomial
5. a

Objective a

7. $(v-3)^2$
9. $(y+4)^2$
11. $(5-n)^2$, or $(n-5)^2$
13. $x(x+10)^2$

Objective b

15. $(t+1)(t-1)$
17. $5a(b+2)(b-2)$
19. $x^2(8y^4+5x^2)(8y^4-5x^2)$

Objective c

21. $(y-z+8)(y-z-8)$
23. $(p+5q+10t)(p+5q-10t)$
25. $\left(\frac{1}{5}+w\right)\left(\frac{1}{5}-w\right)$
27. $(w-1+5p)(w-1-5p)$
29. $5(w-y)^2(w+y)$
31. $(2+x+y)(2-x-y)$
33. $(8+w+t)(8-w-t)$

Objective d

35. $(4t-1)(16t^2+4t+1)$
37. $(z+w)(z^2-zw+w^2)$
39. $(d-0.2)(d^2+0.2d+0.04)$
41. $\left(x-\frac{1}{3}\right)\left(x^2+\frac{1}{3}x+\frac{1}{9}\right)$
43. $10(5t^2+s^2)(25t^4-5t^2s^2+s^4)$
45. $(x+1)(x^2-x+1)(x-1)(x^2+x+1)$
47. $x^6(x^3+y^4)(x^6-x^3y^4+y^8)$

49. $(r+2)(r^2-2r+4)(r-2)(r^2+2r+4)$

Section 4.7

Key Terms

1. common factor
3. grouping

Objective a

5. $(2x+3)(x-2)$
7. $a(a-6)^2$
9. $(x-2)(x+3)(x-3)$
11. $2t(t+5)(3t-2)$
13. $-3y^3(y-4)^2$
15. $3t^2(t^2+4)(t+2)(t-2)$
17. $c^2(x+y)$
19. $(5x-y)(2x-1)$
21. $(t-2+10v)(t-2-10v)$
23. $(3p-2q)^2$
25. $(y+2)(y^2-2y+4)(y-2)(y^2+2y+4)$
27. $(4m-3n)(5m+2n)$
29. $(a^2+b^3+5)(a^2-b^3-5)$
31. $\left(\frac{1}{3}a+\frac{2}{5}\right)^2$

Section 4.8

Key Terms

1. right triangle
3. leg

Objective a

5. $0, 9$
7. $-2, 0$
9. $-2, 8$
11. $0, 12$
13. $-\frac{5}{3}, 1$
15. $-\frac{5}{2}, \frac{3}{8}$
17. $\left\{x \middle| x \text{ is a real number } \textit{and } x \neq 0 \textit{ and } x \neq \frac{1}{7}\right\}$, or $(-\infty, 0) \cup \left(0, \frac{1}{7}\right) \cup \left(\frac{1}{7}, \infty\right)$

Objective b

19. Length: 12 in.; width: 9 in.
21. 10 ft
23. Distance d: 16 ft; pole height: 30 ft
25. Length: 80 ft; width: 60 ft
27. Length: 14 m; width: 6 m

Chapter 5 RATIONAL EXPRESSIONS, EQUATIONS AND FUNCTIONS

Section 5.1

Key Terms

1. simplify

3. invert

Objective a

5. –7, 5
7. $\{x|x \text{ is a real number } \textit{and } x \neq 1\}$, or $(-\infty,1)\cup(1,\infty)$
9. $\{x|x \text{ is a real number } \textit{and } x \neq -4 \textit{ and } x \neq 4\}$, or $(-\infty,-4)\cup(-4,4)\cup(4,\infty)$

Objective b

11. $\dfrac{(y-6)(y-7)}{(y+1)(y-7)}$

Objective c

13. $\dfrac{5a^7b}{2}$

15. $\dfrac{m-5}{m}$

17. $\dfrac{a+2}{a+3}$

Objective d

19. $\dfrac{16}{3t^2}$

21. $\dfrac{(x^2+1)(x+1)}{(x-2)(x-3)(x-1)}$

23. $\dfrac{2(y+3)(y-3)}{(y-8)(y-1)}$

Objective e

25. $\dfrac{1}{x^3y^2}$

27. $\dfrac{5(a-7)}{a-3}$

29. $\dfrac{(x+2)(x+3)}{3(x-5)}$

31. $(k-3)^2$

Section 5.2

Key Terms

1. least common multiple

Objective a

3. 120
5. 60
7. $\frac{11}{20}$
9. $\frac{63}{100}$
11. $60x^2y^4$
13. $m(m+3)^2(m-3)$
15. $12y^3(y-3)^2(y+5)$

Objective b

17. $y+11$
19. $\frac{2t+3}{t-4}$
21. $\frac{7x-41}{(x-7)(x-3)(x+1)}$
23. $\frac{x-3}{(x+3)(x-1)}$

Objective c

25. $\frac{-x-1}{4x-1}$
27. $\frac{36}{y(y-6)}$

Section 5.3

Key Terms

1. quotient
3. dividend

Objective a

5. $7-t^4+3t^5$
7. $-3y^3+y+\frac{6}{y}$
9. $-6s+10s^3t-7t^2$

Objective b

11. $p-1+\frac{9}{p-3}$
13. $2x^2+3x+1+\frac{8x-3}{x^2-3}$

Objective c

15. $a^2 - a - 1 + \dfrac{4}{a+1}$

17. $x^2 - 4x + 16$

19. $x^2 - 6x + 12 + \dfrac{-13}{x+2}$

Section 5.4

Objective a

1. $-\dfrac{13}{10}$

3. 3

5. $\dfrac{p}{1-p}$

7. $\dfrac{3z-3y}{yz}$

9. $\dfrac{-7(n^2-3)(n^2+3)}{10n^2(n^2+9)}$

11. 1

13. $\dfrac{(x-2)(x-5)}{(x+3)(x+1)}$

15. $\dfrac{(x+1)(x+3)}{(x+2)(x+5)}$

Section 5.5

Key Terms

1. rational equation

3. LCD

Objective a

5. 20

7. –6

9. 2, 5

11. 3

13. $-2, 7$

15. 4

17. 7

19. $\dfrac{1}{2}, 3$

Section 5.6

Key Terms

1. $\dfrac{t}{a} + \dfrac{t}{b} = 1$

Objective a

3. $3\frac{3}{4}$ hr

5. 144 min, or 2 hr 24 min

7. $40\frac{10}{11}$ min

Objective b

9. 96 pages

11. $4\frac{1}{2}$ cups

Objective c

13. 10 mph

15. 6 km/h

Section 5.7

Objective a

1. $Y = \dfrac{mt}{t+m}$

3. $p = \pm\sqrt{rw}$

5. $\dfrac{g \pm \sqrt{(-g)^2 + 4h}}{2}$

7. $r = \dfrac{En - EI}{In}$

9. $t_2 = \dfrac{Smt_1 - H}{Sm}$

Section 5.8

Key Terms

1. direct

3. inverse

Objective a

5. 11; $y = 11x$

Objective b

7. $16\frac{2}{3}$ cm

9. 25 min

Objective c

11. $6;\ y=\frac{6}{x}$

Objective d

13. 120 m

Objective e

15. $y=\frac{18}{x^2}$

17. $y=\frac{11}{10}xz^2$

Objective f

19. 3.125 W/m^2

Chapter 6 RADICAL EXPRESSIONS, EQUATIONS, AND FUNCTIONS

Section 6.1

Key Terms

1. square
3. irrational
5. cube

Objective a

7. $8, -8$
9. $-\frac{15}{13}$
11. 0.3
13. 23.707
15. –27.619
17. $\frac{2s}{3t}$
19. Does not exist; 1; $\sqrt{10}$; $\sqrt{6}$
21.

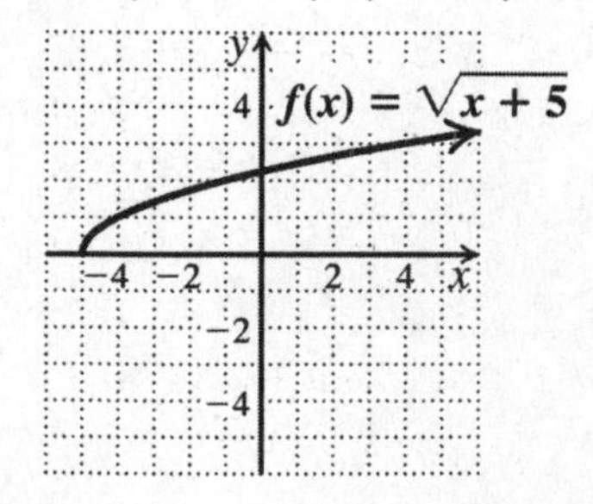

23.

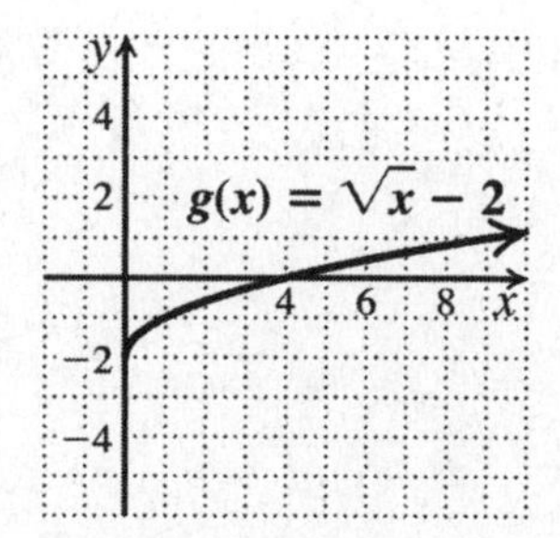

Objective b

25. $|y-5|$

27. $|2x-7|$

Objective c

29. $-5t$

31. Does not exist; 1; does not exist; 2

Objective d

33. $-\frac{2}{3}$

35. $3ab$

Section 6.2

Key Terms

1. c
3. d
5. f

Objective a

7. $\sqrt[4]{a}$
9. $\sqrt{pqr}$
11. 32
13. $10^{1/5}$
15. $(ab^3)^{1/4}$

Objective b

17. $\frac{1}{(3x)^{1/5}}$

19. $3x^2y^{4/5}$

Objective c

21. $3^{1/2}$

23. $x^{6/5}$

Objective d

25. x^5
27. $\sqrt[10]{m}$
29. $\sqrt[6]{7776}$
31. a^9b^6

Section 6.3

Key Terms

1. radicands
3. square

Objective a

5. $10\sqrt{2}$
7. $3x^2\sqrt[3]{y^2}$
9. $x^2y^5\sqrt{x}$
11. $-2cd^3\sqrt[5]{10c^3d^2}$
13. $\sqrt{110}$
15. $7\sqrt{6}$
17. $28x^5$
19. $cd^3\sqrt[3]{c^2d}$

Objective b

21. $\sqrt{6}$
23. $a^2\sqrt{3b}$
25. $\frac{9}{7}$
27. $\frac{4p^2\sqrt{p}}{q^5}$
29. $\frac{a^2b}{c^3}\sqrt[4]{\frac{b^3}{c^2}}$

Section 6.4

Key Terms

1. like radicals

Objective a

3. $-3\sqrt{11}$
5. $3\sqrt{3}+10\sqrt[3]{5}$
7. $(2a^2+9a)\sqrt[3]{5a}$
9. $(2-z)\sqrt{4z-1}$

Objective b

11. $6+10\sqrt[3]{5}$
13. $-6-3\sqrt{10}$
15. $2x+7-2\sqrt{14x}$
17. $\sqrt[4]{72}+\sqrt[4]{225}-\sqrt[4]{24}-\sqrt[4]{75}$

Section 6.5

Objective a

1. $\dfrac{3\sqrt{14}}{10}$

3. $\dfrac{\sqrt[3]{10x^2y}}{xy}$

5. $\dfrac{\sqrt{21}}{7}$

7. $\dfrac{\sqrt[3]{245}}{7}$

Objective b

9. $\dfrac{5-\sqrt{3}}{11}$

11. $\dfrac{p+\sqrt{pq}}{p-q}$

Section 6.6

Key Terms

1. radical equation

Objective a

3. $\dfrac{32}{3}$

5. 17

7. 61

9. 0, 49

11. -1000

Objective b

13. 2

15. 3, 7

17. $\dfrac{3}{4}, 1$

Objective c

19. 592 ft

21. 180 ft

23. About 3.6 ft

Section 6.7

Objective a

1. $\sqrt{65}$, 8.062

3. $\sqrt{65}$, 8.062

5. $\sqrt{22}$, 4.690
7. $\sqrt{180}$ ft, 13.416 ft
9. 32 in.

Section 6.8

Key Terms

1. i
3. complex

Objective a

5. $7i$
7. $10i\sqrt{2}$, or $10\sqrt{2}i$
9. $8-5i\sqrt{6}$, or $8-5\sqrt{6}i$

Objective b

11. $12+3i$
13. $-3-6i$

Objective c

15. -88
17. $8+10i$
19. $-28+3i$
21. $-33-56i$

Objective d

23. 1
25. $-8i$

Objective e

27. $\frac{6}{5}-\frac{3}{5}i$
29. $-\frac{5}{2}-3i$

Objective f

31. Yes
33. No

Chapter 7 QUADRATIC EQUATIONS AND FUNCTIONS

Section 7.1

Key Terms

1. quadratic
3. complete

Objective a

5. a) $\frac{9}{2}i, -\frac{9}{2}i$, or $\pm\frac{9}{2}i$; b) no x-intercepts

7. $\pm\frac{9}{2}i$

9. $5\pm\sqrt{2}$; 3.586, 6.414

11. $-\frac{2}{3}\pm\frac{2\sqrt{2}}{3}$; -1.609, 0.276

Objective b

13. 2,4

15. $-3\pm\sqrt{11}$

17. $\left(2-\sqrt{2},0\right),\left(2+\sqrt{2},0\right)$

19. $\left(\frac{1}{4}-\frac{\sqrt{5}}{4},0\right),\left(\frac{1}{4}+\frac{\sqrt{5}}{4},0\right)$

21. $-\frac{1}{3}\pm\frac{\sqrt{13}}{3}$

23. $-4\pm 3i$

Objective c

25. About 7.8 sec

Section 7.2

Key Terms

1. $\pm\sqrt{k}$

Objective a

3. $-\frac{3}{2}\pm\frac{\sqrt{17}}{2}$

5. $\frac{1}{2}\pm\frac{\sqrt{11}}{2}i$

7. $-2\pm\frac{3\sqrt{2}}{2}$

9. $-\frac{1}{4}\pm\frac{\sqrt{105}}{20}$

11. 1, 3

13. $3\pm 4i$

15. $-1\pm\sqrt{6}$; -3.449, 1.449

17. $\frac{-2\pm 3\sqrt{2}}{2}$; -3.121, 1.121

Section 7.3

Objective a

1. Length: 9 ft; width: 4 ft

3. Base: $2\sqrt{37}-2$ ft; height: $2+2\sqrt{37}$ ft

5. 76 and 77

7. First part: 50 mph; second part: 45 mph

9. Tim: 300 mph, David: 375 mph; or Tim: 125 mph, David 200 mph

11. About 14.4 mph

13. $x=\dfrac{-5+\sqrt{25+12y}}{6}$

15. $q=\dfrac{25p}{M^2}$

17. $v=\dfrac{-4x_0+2\sqrt{4x_0^{\ 2}+2at}}{a}$

19. $d=\dfrac{H^2}{2.56}$

Section 7.4

Key Terms

1. discriminant

3. conjugates

Objective a

5. Two real

7. Two real

9. Two nonreal

11. Two real

Objective b

13. $x^2-8x+16=0$

15. $5x^2-17x-12=0$

17. $x^2-90=0$

Objective c

19. $\pm1,\ \pm4$

21. 25

23. $-\frac{1}{6},\frac{1}{5}$

25. -1, 1024

27. $(-2,0),(-1,0),(3,0),(4,0)$

Section 7.5

Key Terms

1. parabola

3. vertex

Objective a, b

5. Vertex: $(-2,0)$;
axis of symmetry: $x=-2$

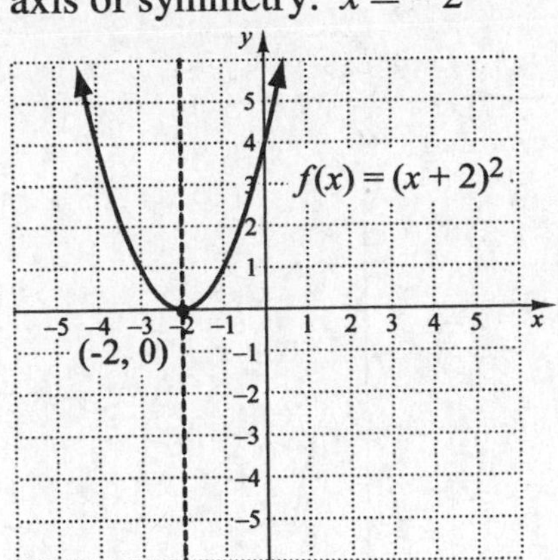

7. Vertex: $(3,0)$;
axis of symmetry: $x=3$

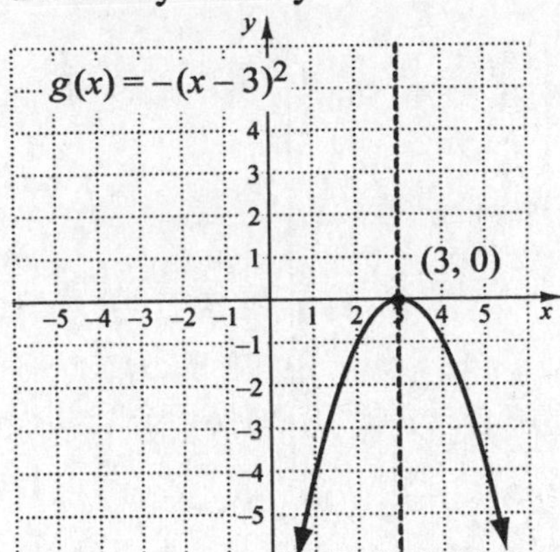

Objective c

9. Vertex: $(-2,-3)$;
axis of symmetry: $x=-2$;
minimum: -3

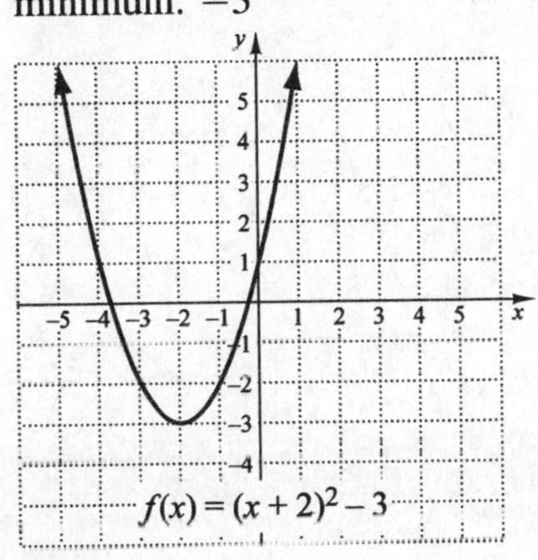

11. Vertex: $(3,4)$;
axis of symmetry: $x=3$;
maximum: 4

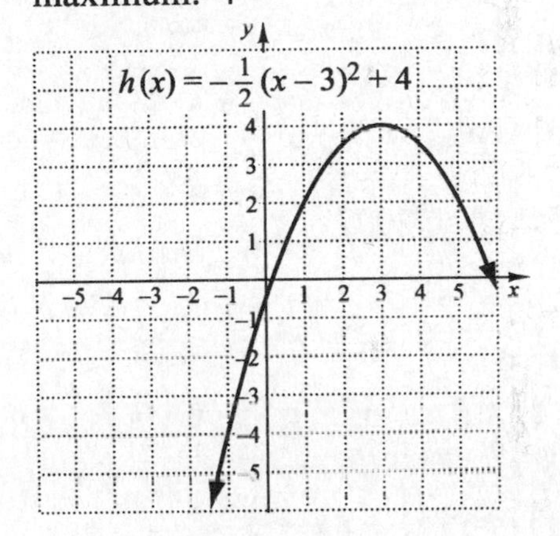

Section 7.6

Key Terms

1. minimum

3. y-intercept

Objective a

5. a) $f(x)=(x+4)^2-21$; b) vertex: $(-4,-21)$; axis of symmetry: $x=-4$

7. a) Vertex: $(2,-5)$;
   axis of symmetry: $x=2$;
   b) minimum: -5
   c)

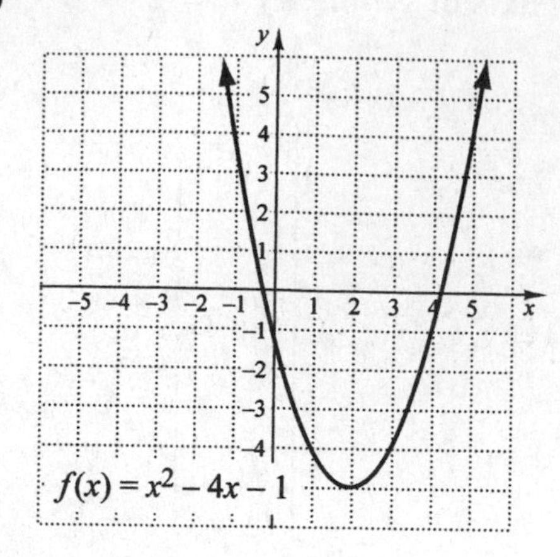

9. a) Vertex: $(1,-4)$;
   axis of symmetry: $x=1$;
   b) maximum: -4
   c)

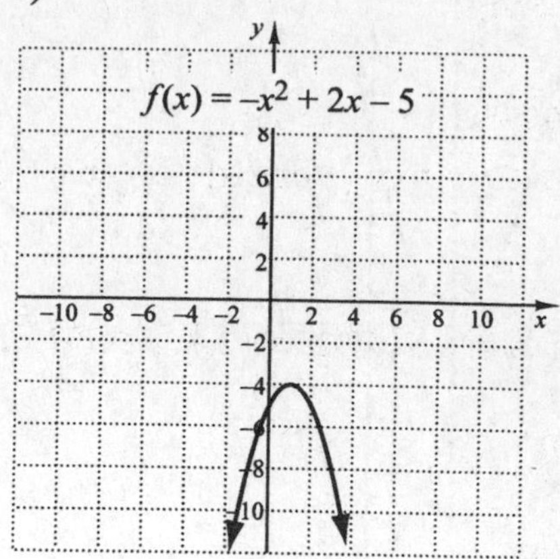

Objective b

11. $(-3-\sqrt{10},0),(-3+\sqrt{10},0);(0,-1)$

13. $(-2,0),(3,0);(0,6)$

Section 7.7

Objective a

1. \$6; March 2007
3. 21 ft by 21 ft
5. 5 ft by 5 ft
7. -25; 5 and -5

Objective b

9. $f(x)=mx+b$
11. Neither quadratic nor linear
13. $f(x)=\frac{5}{3}x^2+2x-\frac{2}{3}$

Section 7.8

Key Terms

1. polynomial
3. test points

Objective a

5. $\left(-1,\frac{9}{2}\right)$

7. $(-\infty,-1]\cup[4,\infty)$, or $\{x|x\leq-1\ or\ x\geq4\}$
9. $\varnothing$

Objective b

11. $(3,\infty)$, or $\{x|\ x>3\}$
13. $[-3,-2)\cup[5,\infty)$, or $\{x|-3\leq x<-2\ or\ x\geq5\}$

Chapter 8 EXPONENTIAL AND LOGARITHMIC FUNCTIONS

Section 8.1

Key Terms

1. exponential
3. decreasing

Objective a

5.

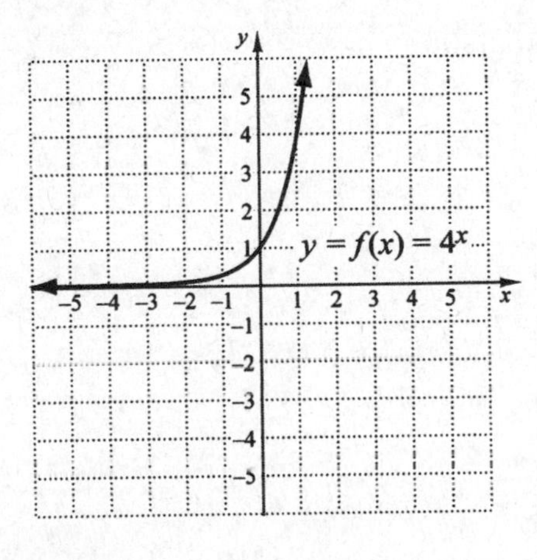

7.

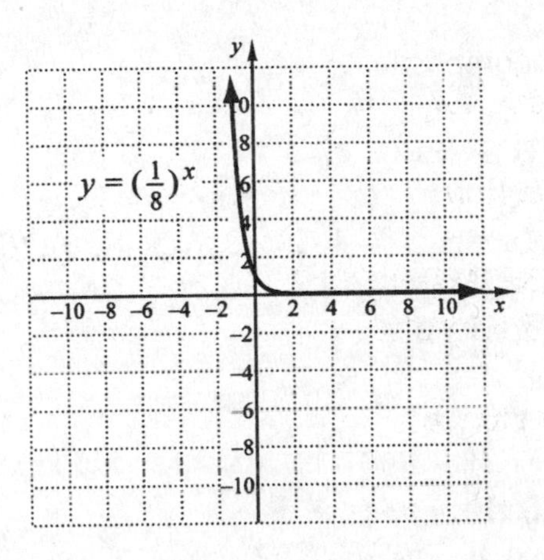

Objective b

9.

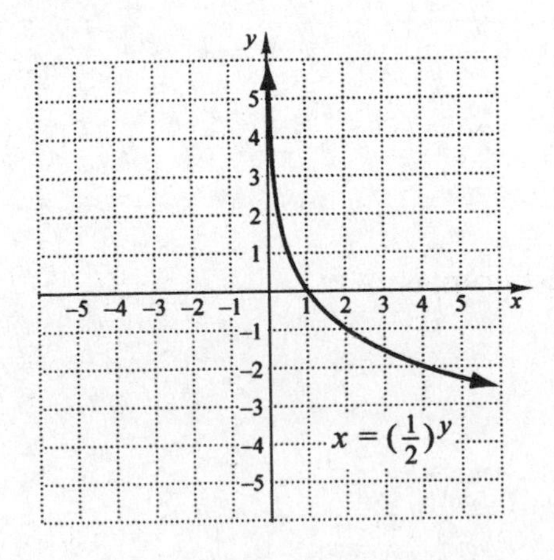

11.

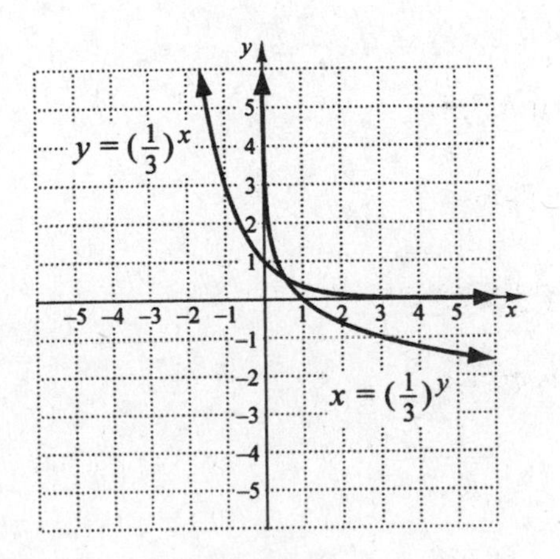

Objective c

13. a) About \$12.4 billion;
about \$798 billion;

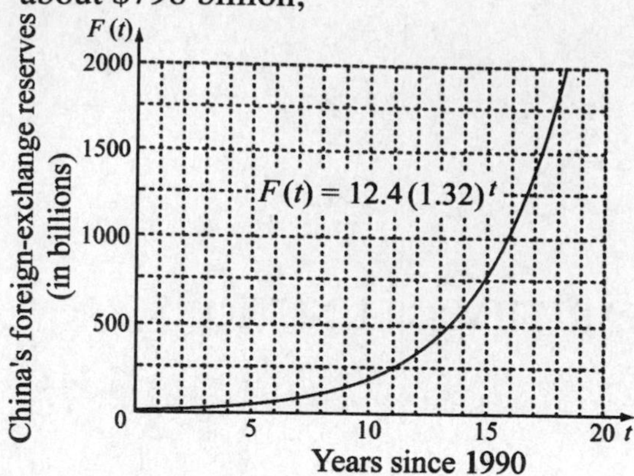

Section 8.2

Key Terms

1. composite
3. inverse

Objective a

5. Inverse: $\{(-1, 2), (3, 6), (-4, 6), (0, 8)\}$

Objective b

7. Yes
9. No

Objective c

11. $f^{-1}(x) = x - 1$
13. Not one-to-one
15. $f^{-1}(x) = \sqrt[3]{x-2}$

17.

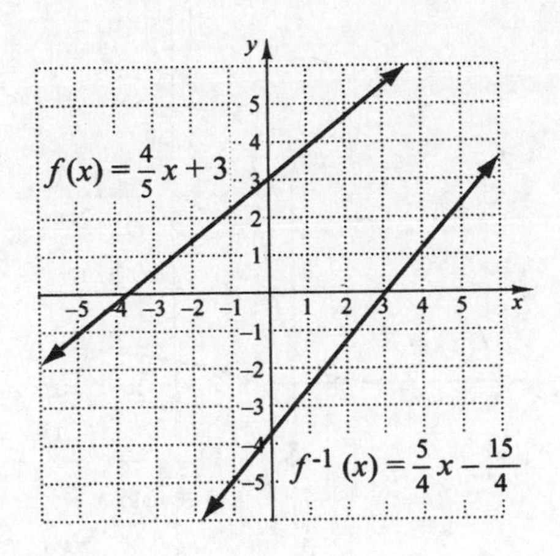

Objective d

19. $-1.38x - 8; -1.38x + 8.22$

21. $\frac{80}{x^2} + 2; \frac{4}{5x^2 + 2}$

23. $f(x) = x^2$; $g(x) = 10 - 3x$

25. $f(x) = \frac{4}{x}$; $g(x) = \sqrt{2x+7}$

Objective e

27. (1) $(f^{-1} \circ f)(x) = f^{-1}(f(x))$

$$= f^{-1}\left(\sqrt[3]{x+2}\right) = \left(\sqrt[3]{x+2}\right)^3 - 2$$

$$= x + 2 - 2 = x;$$

(2) $(f \circ f^{-1})(x) = f(f^{-1}(x))$

$$= f\left(x^3 - 2\right) = \sqrt[3]{x^3 - 2 + 2}$$

$$= \sqrt[3]{x^3} = x$$

29. $f^{-1}(x) = \frac{1}{4}x$

31. $f^{-1}(x) = \frac{1}{x}$

33. a) 5; b) yes; $f^{-1}(x) = x + 32$; c) 40

Section 8.3

Key Terms

1. logarithmic

3. common

Objective a

5.

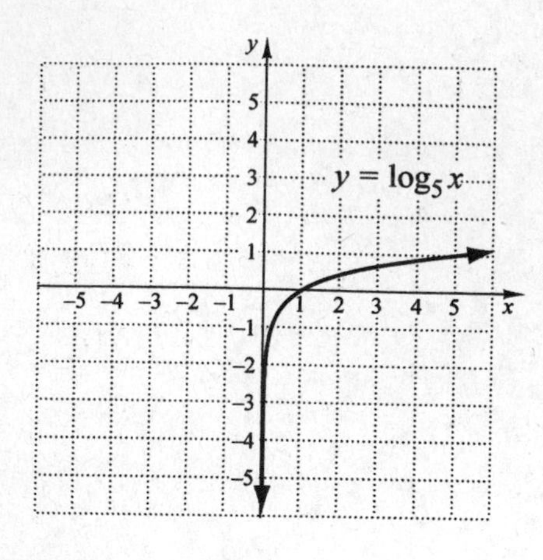

7.

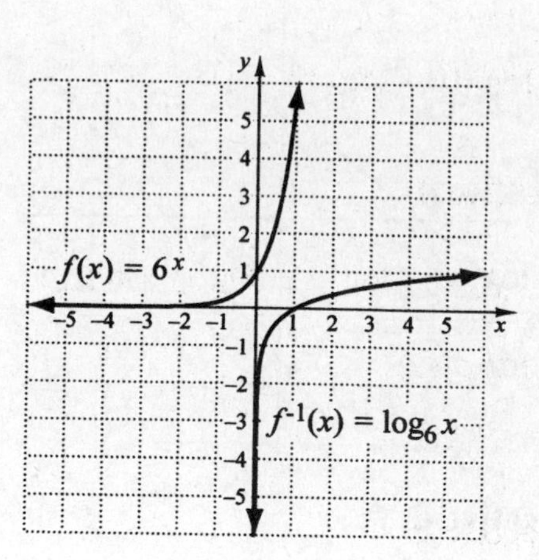

Objective b

9. $\frac{2}{3} = \log_{27} 9$

11. $-3 = \log_e 0.0498$

13. $a^p = t$

Objective c

15. 16

17. 1000

19. 5

21. –1

23. 0

Objective d

25. –0.5376

27. –0.3826

Section 8.4

Key Terms

1. c

3. e

5. f

Objective a

7. $\log_2 4 + \log_2 128$

9. $\log_a (9 \cdot 10)$, or $\log_a 90$

Objective b

11. $6\log_a t$

Objective c

13. $\log_3 10 - \log_3 13$

15. $\log_a \frac{25}{11}$

Objective d

17. $\log_b w + \log_b x + \log_b y + \log_b z$

19. $3\log_a w + \log_a x - \log_a y - 2\log_a z$

21. $\log_x p^5 q^4$

23. $\log_a \frac{3y^6}{z^5}$

25. 1.183

27. −0.356

Objective e

29. 19

31. 6

Section 8.5

Key Terms

1. natural

3. domain

Objective a

5. 2.0794

7. 5.1160

9. 3

11. −5.2591

13. 1.7986

Objective b

15. 2.8614

17. 3.0276

Objective c

19. Domain: $\mathbb{R}$; range: $(0, \infty)$

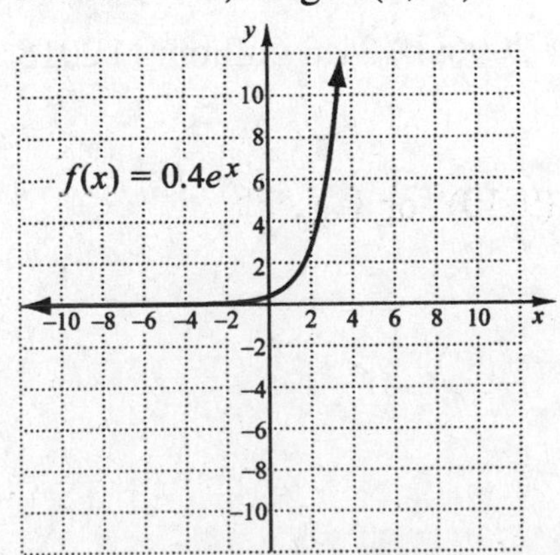

21. Domain: $(2, \infty)$; range: $\mathbb{R}$

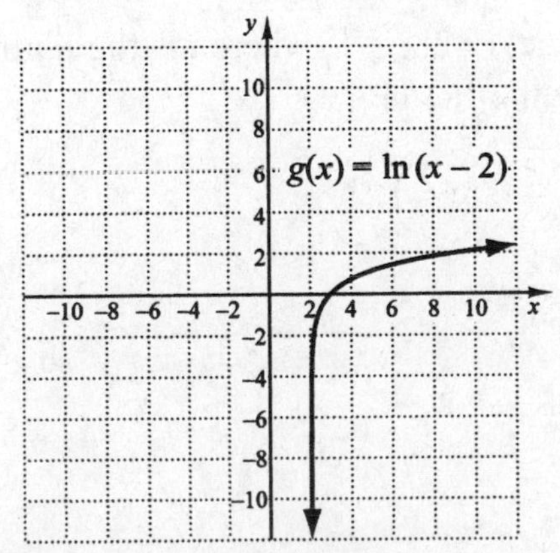

Section 8.6

Key Terms

1. exponential
3. exponential

Objective a

5. 3.4594
7. 4
9. 0.3054
11. 2
13. 48.1589

Objective b

15. $\frac{1}{9}$
17. $e^9 \approx 8103.084$
19. $10^{3/2} \approx 31.623$
21. 2
23. $\frac{251}{62}$

Section 8.7

Key Terms

1. decibel
3. growth
5. decay

Objective a

7. 98 dB
9. 10.5
11. a) 72%; b) 63%; c) 55%

Objective b

13. a) 2.7 yr; b) 17.7 yr
15. a) $P(t) = 28e^{0.03t}$, where t is the number of years after 2006; b) 31.6 million; c) 2018
17. About 1354 yr

Chapter 9 CONIC SECTIONS

Section 9.1

Key Terms

1. parabola

3. center

Objective a

5.

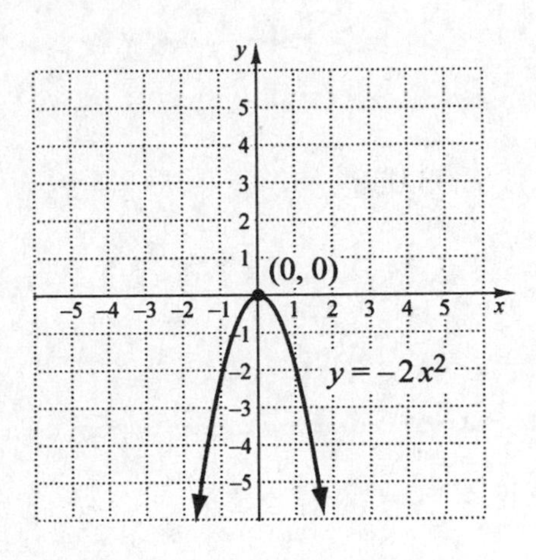

7.

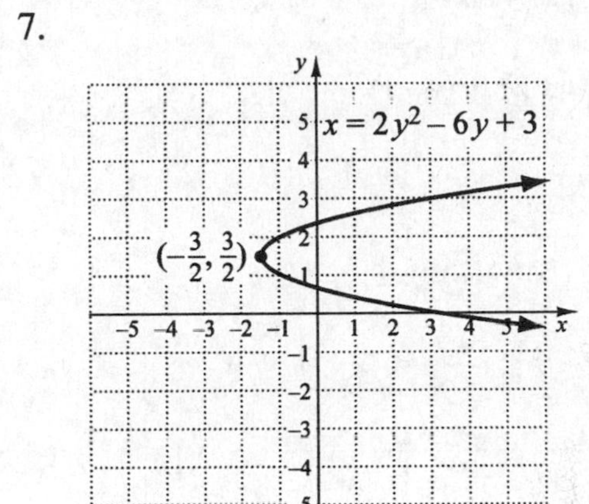

Objective b

9. 10

11. $\dfrac{\sqrt{5}}{2} \approx 1.118$

13. $\sqrt{53} \approx 7.280$

Objective c

15. $(5,-7)$

17. $\left(\dfrac{\sqrt{3}+\sqrt{7}}{2}, \dfrac{5}{2}\right)$

Objective d

19. $(-2,1)$; 2

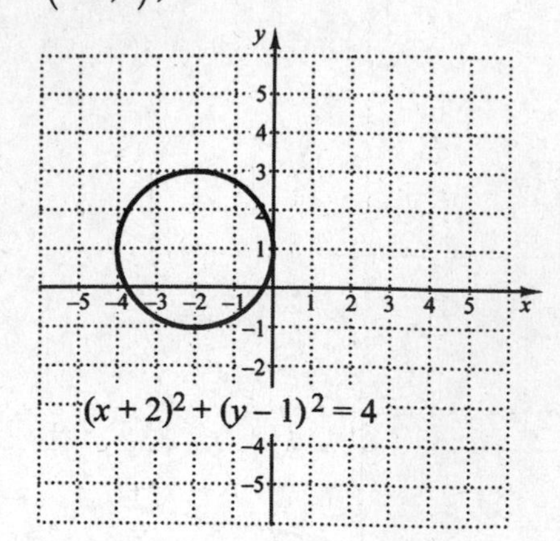

21. $(0,0)$; $\sqrt{5}$

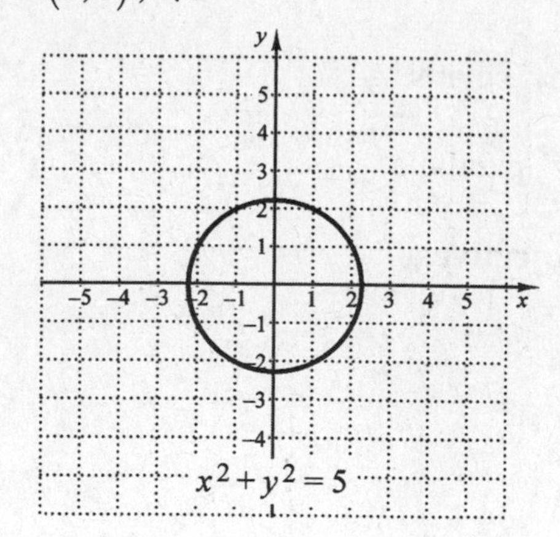

23. $(x-6)^2+(y-2)^2=6$

25. $(x+2)^2+(y+5)^2=245$

27. $\left(0,-\frac{5}{2}\right)$; $\frac{\sqrt{37}}{2}$

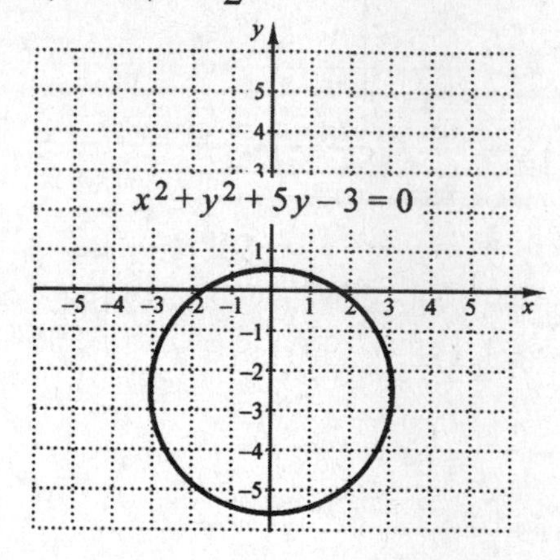

Section 9.2

Key Terms

1. ellipse

3. horizontal

Objective a

5.

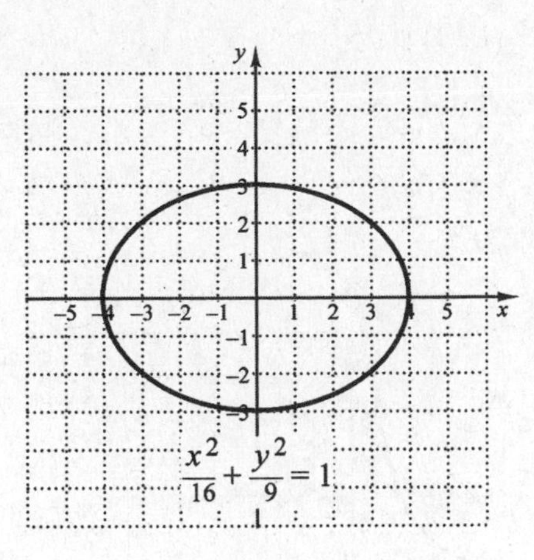

7.

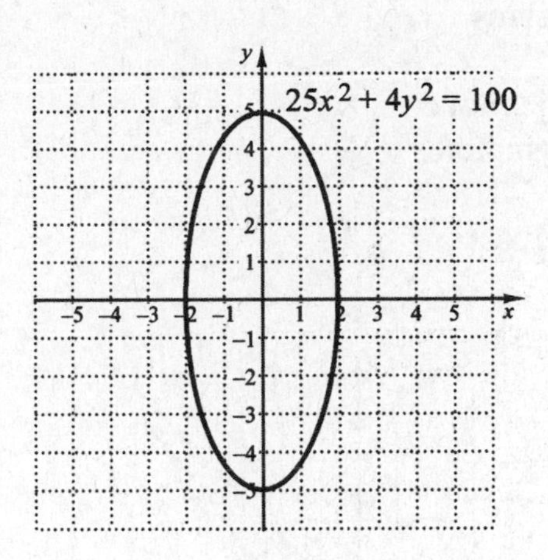

9.

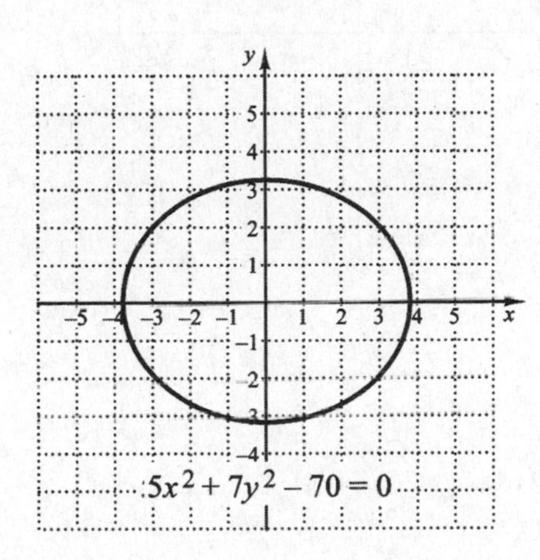

11.

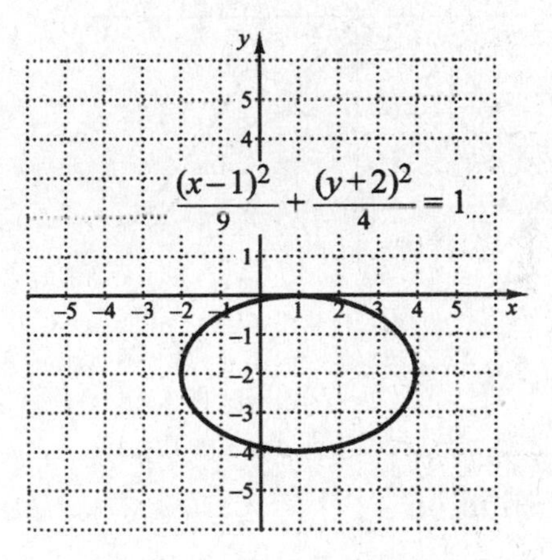

13.

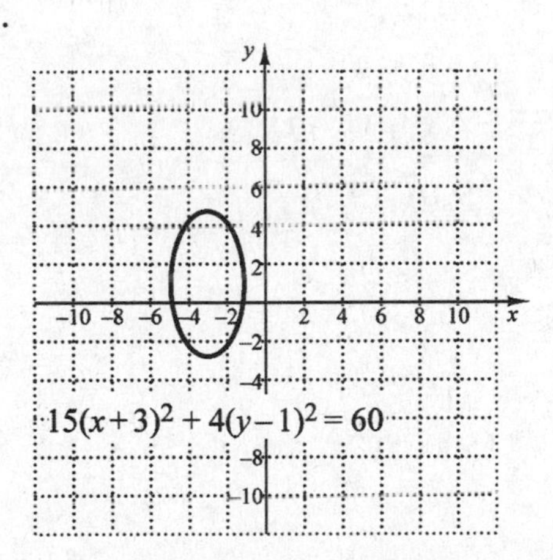

Section 9.3

Key Terms

1. hyperbola
3. horizontal
5. asymptotes

Objective a

7.

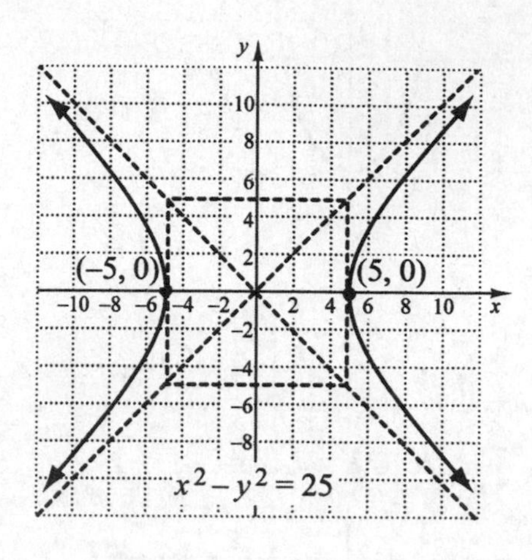

Objective b

9.

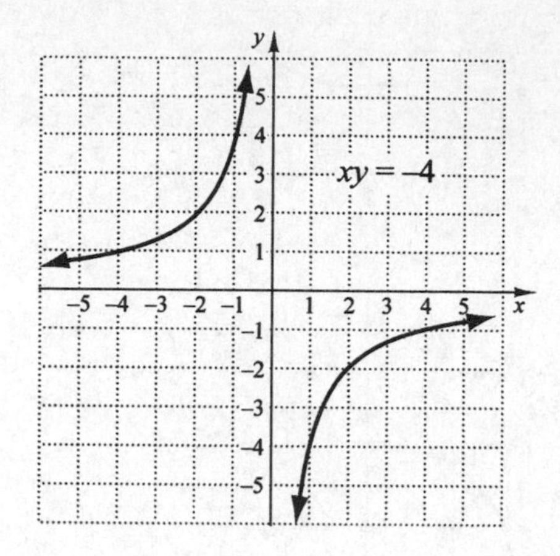

Section 9.4

Key Terms

1. substitution

Objective a

3. $(-3,-4)$, $(4,3)$
5. $\left(-2,\frac{1}{2}\right)$, $(1,-1)$
7. $(1,4)$, $(4,1)$
9. $(-9,0)$, $(9,0)$
11. $(1,\sqrt{3})$, $(1,-\sqrt{3})$, $(-1,\sqrt{3})$, $(-1,-\sqrt{3})$

Objective b

13. Length: 5 ft; width: 1 ft
15. $1500, 5%